Experimenting With Industrial Electrical Systems

by
Dale R. Patrick, Professor
and
Stephen W. Fardo, Associate Professor
Department of Industrial Education and Technology
College of Applied Arts and Technology
Eastern Kentucky University

Howard W. Sams & Co., Inc.
4300 WEST 62ND ST. INDIANAPOLIS, INDIANA 46268 USA

FIRST EDITION
FIRST PRINTING—1978

International Standard Book Number: 0-672-21473-3
Library of Congress Catalog Card Number: 77-083975

Printed in the United States of America.

Preface

This laboratory-oriented activities manual is intended for use by vocational/technical or college-level industrial electrical programs. These learning activities are organized to parallel the content of *Industrial Electrical Systems*, by Dale R. Patrick and Stephen W. Fardo (also published by Howard W. Sams & Co., Inc.). In this book, Unit 1 parallels Chapter 1 in the textbook, Unit 2 parallels Chapter 2, etc. The experiments are presented by using a "single-concept" approach. The individual experiments require a brief time to construct and to make the necessary measurements and observations. By utilizing the "single-concept" approach, student learning is facilitated. Circuit application and operation is stressed in each learning activity.

The learning activities supplement the content of *Industrial Electrical Systems*, or a similar text, through an experimental approach. Each learning activity is organized as follows:

1-1
(First activity of Unit 1.)
TITLE
(Topic of the activity.)

Objective: An overview of the activity in relation to some operational system; also states the purpose of the activity and suggests observations.

Equipment: Necessary laboratory equipment and materials.

Procedure: Logical step-by-step sequence for completing the learning activity; subdivided into "sections" as necessary. Maximum use is made of illustrative charts which aid in the recording of laboratory data.

Analysis: Specific related questions and problems that supplement the laboratory activity.

The activities of Unit 1 are primarily included as a review of basic electrical theory. Most of these activities do not require any laboratory equipment. The completion time for these activities is relatively short, providing the instructor and student with a rapid means of assessing basic electrical problem-solving skills and knowledge of equipment operation.

The activities of Unit 2 provide a specific working knowledge of codes, symbols, and standards for industrial systems. No laboratory equipment is required for these activities; however, each activity is directed toward important industrial guidelines that industrial technicians must understand, such as the use of the National Electrical Code, conducting an actual OSHA inspection, or using industrial standards.

Unit 3 contains laboratory activities that provide an in-depth understanding of electrical energy sources. Alternating-current sources are mainly single-phase and three-phase alternators. Direct-current sources include primary and secondary chemical cells, mechanical generators, and rectification systems. Direct current for industrial use is produced mainly by rectification. Therefore, a thorough coverage of single-phase and three-phase rectification, filtering systems, regulation, and voltage multiplication methods is included in this unit. Semiconductor circuits are primarily investigated, with inclusion of some vacuum-tube and gaseous-tube circuits.

Power distribution systems are studied in Unit 4. The unit centers around single-phase and three-phase transformer operation and applications. Conductors and insulators are also examined, along with printed circuit construction and soldering and fastening exercises.

Unit 5 begins with the basics of single-phase industrial electrical load circuits (RL, RC, RCL, series and parallel), with a unique, easy-to-understand laboratory approach used to simplify the voltage, current, power, and impedance relationships of these circuits. Three-phase balanced and unbalanced loads are studied, along with basic electrical lighting and heating loads. The main content of this unit deals with mechanical loads (motors). Since motors are the workhorses of industry, they deserve considerable lab attention, particularly from an applications point of view. The activities which investigate the operation, application, and troubleshooting of various types of dc motors, and single-phase and three-phase ac motors are organized so that either commercially manufactured or laboratory constructed machines may be used. Several methods of motor control are also included as separate laboratory activities.

Unit 6 provides a basic understanding of indicating systems. Measurement of various quantities is very important for all industries. Several measurement methods, such as multifunction meters, the Wheatstone bridge, oscilloscopes, chart recorders, power measuring methods, and telemetering systems are included. Some basic transducing systems used to measure physical quantities, such as thermocouples, strain gauges, and thermistors, are investigated in this unit.

The last unit, Unit 7, provides several activities which illustrate industrial applications of hydraulic and pneumatic fluid power systems.

The overall organization of these activities provides the student with a broad overview of industrial systems from the basic level through some rather sophisticated circuits and systems. Upon completion of these activities, or as many as may be completed due to the availability of lab equipment, the student should possess an excellent operational understanding of industrial systems.

DALE R. PATRICK
STEPHEN W. FARDO

Laboratory Equipment and Supplies

The following list of equipment and supplies is suggested for each laboratory area to complete the activities included in this manual:

Resistors

10 Ω, 20 W
27 Ω, 10 W
50 Ω, 20 W
100 Ω, 1 W (2)
100 Ω, 10 W
250 Ω, 1 W
250 Ω, 25 W
270 Ω, 2 W
470 Ω, 1 W
500 Ω, 2 W (2)
500 Ω, 10 W
560 Ω, 1 W
680 Ω, 1 W
760 Ω, 1 W
1000 Ω, ½ W
1000 Ω, 2 W (3)
1000 Ω, 10 W
1200 Ω, 10 W
2500 Ω, 10 W
4700 Ω, 1 W
5000 Ω, 2 W
5000 Ω, 10 W
6800 Ω, 1 W
8200 Ω, 1 W
10,000 Ω, 2 W (2)
10,000 Ω, 10 W
15,000 Ω, 2 W
47,000 Ω, 1 W
68,000 Ω, 1 W
Decade resistance box

Capacitors

100 pF, 25 V dcw
0.1 μF, 50 V dcw
1 μF, 100 V dcw (3)
2 μF, 100 V dcw
3.75 μF, 330 V ac
4 μF, 25 V dcw
10 μF, 100 V dcw
25 μF, 50 V dcw
30 μF, 350 V dcw
40 μF, 100 V dcw
500 μF, 50 V dcw

Semiconductors

Diode—1N4004 (4)
Zener diode—1N1775
Transistor—2N1184 (2)
Silicon controlled rectifier—GE-C22B (3)
Integrated circuit—μA723C
Diac—GE-X13
Triac—GE-X12
Field-effect transistor—GE-FET-1
Timer—555 (2)
Transistor—2N3053 (2)

Transformers and Inductors

Inductor—8 H (3)
Center-tapped transformer—120 V ac primary, 6.3/12.6 V ac secondary
Center-tapped transformer—120 V ac primary, 2.5 V ac secondary
Center-tapped transformer—120 V ac primary, 320 V ac secondary
Power transformer, with multiple secondary windings
Transformer—2:1 step-down (3)
Dissectible transformer, with 500-turn, 100-turn, and 1000-turn windings
Inductor—15 H

Rotating Machinery (Motors and Generators)*

Three-phase ac generator
Variable-speed prime mover
Tachometer—0 to 100,00 r/min
Permanent-magnet dc generator/motor
Separately excited dc generator
Series-wound dc generator/motor (universal motor)
Shunt-wound dc generator/motor
Ac split-phase motor
Capacitor-start ac motor
Ac shaded-pole motor
Three-phase ac induction motor
Three-phase ac synchronous motor
Motor dynamometer unit, with an assortment of ¼- to ½-horsepower motors
Synchro generator/motor set
Ac synchronous motor—Superior Electric Type SS150
Dc stepping motor—Superior Electric Type M061-FD02
Single-phase ac alternator

Miscellaneous Equipment

Lamp—40 W
Lamp—60 W (3)
Lamp—100 W (2)
Porcelain lamp sockets (3)
Vacuum tube—6X4
Potentiometer—10,000 Ω
Potentiometer—5000 Ω (3)
Potentiometer—1000 Ω
Potentiometer—500,000 Ω
Potentiometer—2500 Ω
Potentiometer—50,000 Ω
Potentiometer—100,000 Ω
Potentiometer—50 Ω
Switch—single-pole, single-throw (3)
Switch—double-pole, double-throw
Switch—push-button, normally open (2)
Switch—push-button, normally closed
American wire gauge
Electrical conductors—six 10-inch samples of various types
Copper-clad printed circuit board
Ferric-chloride printed circuit board etching solution
Glass tray—for printed circuit board etching
Steel wool
Carbon paper
Acid resist pen
Acid resist tape
Acid resist paint
Fluorescent lamp
Fluorescent-lamp ballast
Fluorescent-lamp starter
Vapor lamp—neon
Vapor lamp—argon
Conical heating coil—660 W
Electromagnetic coil, with open center area of 2-inch diameter
Soft iron core—1-inch diameter, 6 inches long
Rheostat—0 to 500 Ω, 20 W
Rheostat—0 to 25 Ω, 40 W
Magnetic contactor—120-V coil, 4 contacts (2)
Meter movement—0 to 100 μA, 500 Ω, zero-centered
Thermistor—JA41J1
Thermistor—KA31L1
Thermistor—GE-X15
Relay—400-Ω coil, spdt
Relay—12-V, 125-Ω coil, Guardian Electric No. 1335-2C-12D
Reed switch—GE-X7

---

* A dissectible rotating machinery unit, such as those available through several commercial distributors, may be substituted for or used to supplement the actual machines, if desired.

Laboratory Equipment and Supplies

Reed switch coil, bobbin wound—General Electric
Reed-switch magnet
Thermocouple—Type J (2)
Soldering iron—25 to 50 W
Solder—rosin core
Desoldering tool
Wire strippers
Side-cutting pliers
Needle-nose pliers
Solderless wire connectors
Crimping tool
Gaseous-tube diodes—866A (2)
Gaseous-tube regulator—0A2
Mercury thermometer

Measuring Equipment and Power Sources

Multifunction meter (volt-ohm-milliammeter)
Ac current meter—0 to 20 A, including ac mA ranges
Variable dc power supply—0 to 150 V
Variable ac power supply—0 to 120 V, single-phase
Variable three-phase power supply—0 to 208 V
Dry cells—1.5 V (3)
Oscilloscope
Electronic multifunction meter (vacuum-tube voltmeter, transistorized voltmeter, or digital voltmeter)
Growler—armature tester
Megger—insulation tester
Chart recorder
Single-phase wattmeter—2000 W or higher (2)
Watt-hour meter

Hydraulic Equipment

Tachometer
Pressure relief valve
Directional control valve—4-way, 3-position
Flow meter
Cylinder—1.125-inch bore, double-acting (2)
Cylinder—1.125-inch bore, single-acting
Cylinder—0.5-inch bore, single-acting
Check valve (2)
Flow control valve—adjustable
Connectors—"T" type
Gauge—160 lb/in^2
Hoses—assortment of quick-disconnect type
Tubing—assortment of transparent pieces
Hydraulic fluid

Pneumatic Equipment

Gauge, air pressure—0 to 100 lb/in^2 (psig) (2)
Air supply—0 to 100 lb/in^2 (psig)
Check valve
Gauge, system pressure—0 to 100 lb/in^2 (psig)
Cylinder—1.5-inch, double-acting
Flow control valve—adjustable
Hoses—assortment of quick-disconnect type
Directional control valve—4-way, 3-position
Pneumatic motor—Gast No. 1AM vane type, reversible
Valve—manual shutoff
Muffler
Connector—"T" type
Platform plate
Weight—10 pounds

Contents

UNIT 1

Review of Basic Electrical Theory

UNIT 2

Codes, Standards, and Symbols for Industrial Systems

UNIT 3

Sources of Electrical Energy

UNIT 4

Power Distribution Systems in Industry

UNIT 5

Industrial Electrical Loads

UNIT 6

Indicating Systems

UNIT 7

Mechanical Power Systems

APPENDIX 1

APPENDIX 2

APPENDIX 3

Resistor Color Code

Objective

In this activity, you will review the resistor color code and show your knowledge by identifying the values of several resistors on a self-test. Study the resistor color code shown in Table 1-1A, then complete the self-test.

Table 1-1A. Resistor Color Code

Color	1st Significant Figure	2nd Significant Figure	Multiplier	Tolerance
Black	0	0	1	
Brown	1	1	10	
Red	2	2	100	
Orange	3	3	1000	
Yellow	4	4	10,000	
Green	5	5	100,000	
Blue	6	6	1,000,000	
Violet	7	7	10,000,000	
Gray	8	8		
White	9	9		
Gold*	—	—	0.1	5%
Silver*	—	—	0.01	10%
No Color	—	—		20%

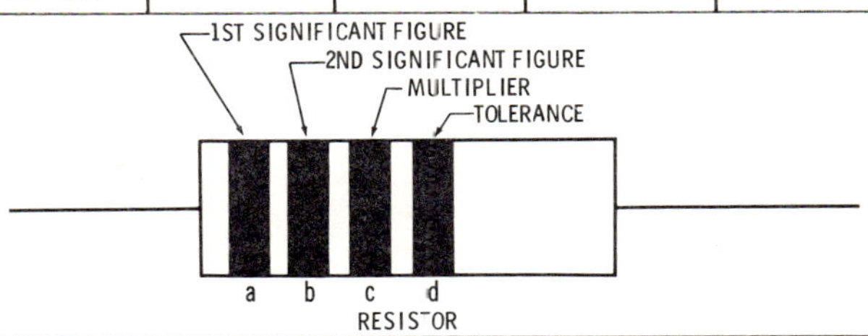

* When resistors have a value of less than 10 ohms, the third color band is a decimal multiplier. The two colors used are: gold = ×0.1 and silver = ×0.01.

Resistor Color Code—Self-Test

For each of the following color codes, identify the resistance value and tolerance of the resistor. Place your answer in the proper blank space.

1. (a) red (b) red (c) blue
   1. Answer: ______ ohms, ______ tolerance.
2. (a) black (b) brown (c) brown (d) gold
   2. Answer: ______ ohms, ______ tolerance.
3. (a) grey (b) red (c) black (d) silver
   3. Answer: ______ ohms, ______ tolerance.
4. (a) white (b) brown (c) orange (d) gold
   4. Answer: ______ ohms, ______ tolerance.
5. (a) yellow (b) violet (c) orange
   5. Answer: ______ ohms, ______ tolerance.
6. (a) brown (b) green (c) orange
   6. Answer: ______ ohms, ______ tolerance.
7. (a) blue (b) gray (c) black (d) gold
   7. Answer: ______ ohms, ______ tolerance.
8. (a) green (b) blue (c) silver (d) gold
   8. Answer: ______ ohms, ______ tolerance.
9. (a) brown (b) black (c) black
   9. Answer: ______ ohms, ______ tolerance.
10. (a) orange (b) white (c) black (d) gold
    10. Answer: ______ ohms, ______ tolerance.
11. (a) red (b) violet (c) gold (d) gold
    11. Answer: ______ ohms, ______ tolerance.
12. (a) brown (b) brown (c) black (d) gold
    12. Answer: ______ ohms, ______ tolerance.

13. (a) violet (b) green (c) orange (d) gold

13. Answer: _______ ohms, _______ tolerance.

14. (a) yellow (b) violet (c) green

14. Answer: _______ ohms, _______ tolerance.

15. (a) green (b) blue (c) orange (d) silver

15. Answer: _______ ohms, _______ tolerance.

16. (a) blue (b) gray (c) orange

16. Answer: _______ ohms, _______ tolerance.

17. (a) yellow (b) orange (c) brown (d) gold

17. Answer: _______ ohms, _______ tolerance.

18. (a) orange (b) orange (c) red (d) gold

18. Answer: _______ ohms, _______ tolerance.

19. (a) brown (b) black (c) green

19. Answer: _______ ohms, _______ tolerance.

20. (a) red (b) red (c) red

20. Answer: _______ ohms, _______ tolerance.

Basic Electrical Symbols

Objective

There are certain basic electrical symbols that are fundamental to the understanding of industrial electrical diagrams. Many of these symbols will be used in the laboratory activities that follow. In this activity, you should study and learn to recognize the basic symbols shown in Table 1-2A, and demonstrate your understanding by completing the self-test.

Table 1-2A. Electrical Symbols

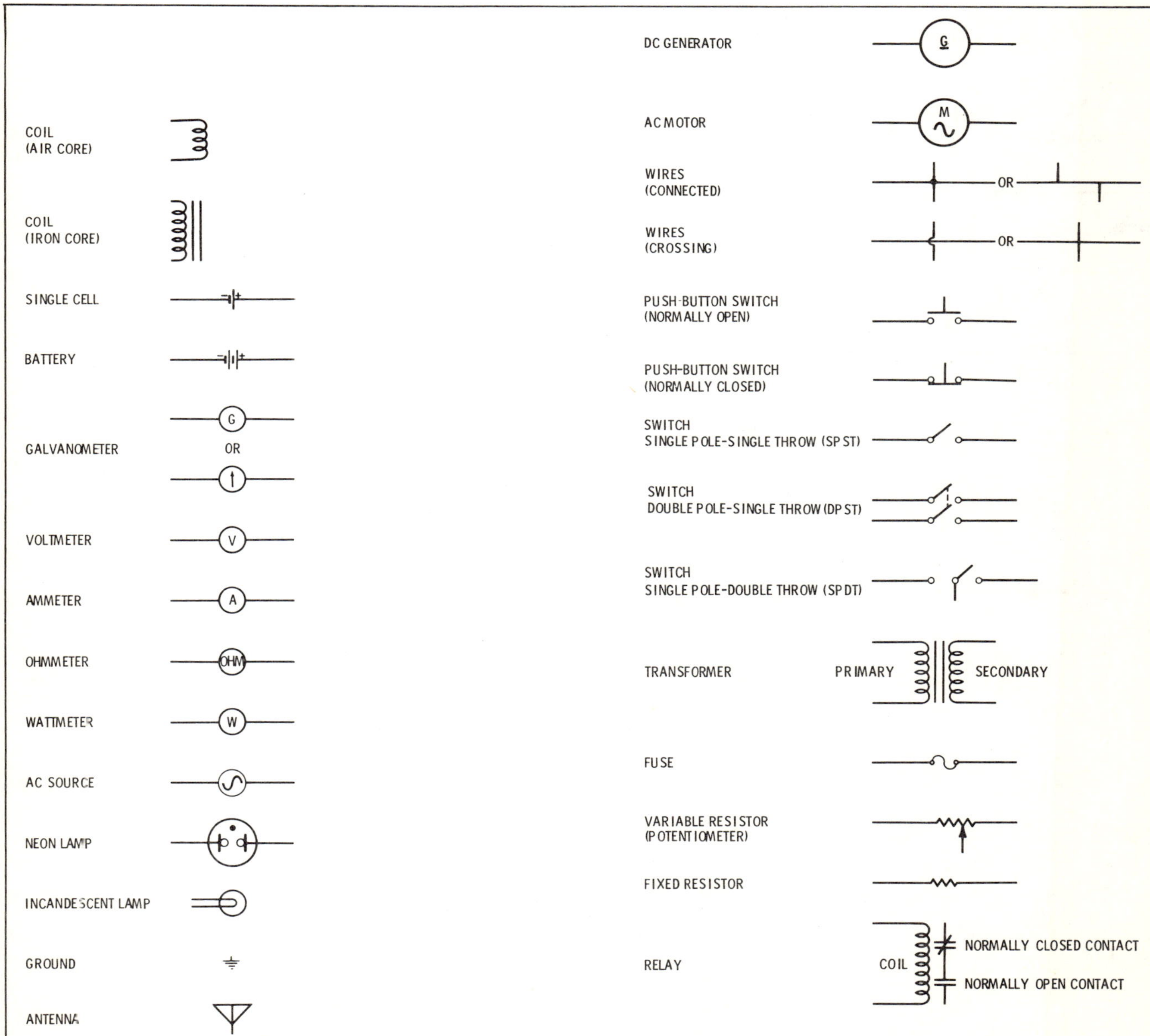

Table 1-2A (cont)

FIXED CAPACITOR	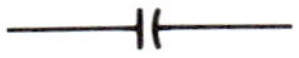	TRANSISTOR (PNP)	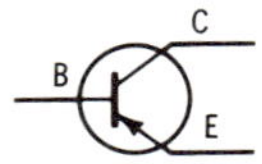	B = BASE C = COLLECTOR E = EMITTER
VARIABLE CAPACITOR		TRANSISTOR (NPN)	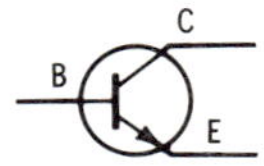	
DIODE (RECTIFIER)		TUBE ELEMENTS:		
DIODE TUBE		FILAMENT		
		CATHODE		
TRIODE TUBE	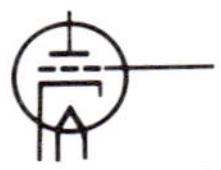	PLATE		
		GRID		

Electrical Symbols—Self-Test

Without looking at the previous symbols, identify each of the symbols that follow. Place the correct response in the space provided.

1.  ________________

2.  ________________

3.  ________________

4.  ________________

5. 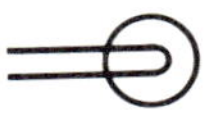 ________________

6.  ________________

7.  ________________

8.  ________________

9.  ________________

10.  ________________

11.  ________________

12.  ________________

Ohm's Law Problems

Objective

A prerequisite for understanding industrial electrical systems is a knowledge of Ohm's law. This fundamental relationship between voltage (V), current (I) and resistance (R) must be understood before electrical concepts become meaningful. In this activity, you will complete some basic practical problems by applying Ohm's law.

Ohm's Law Problems—Self-Test

1. A doorbell requires 0.2 ampere of current in order to ring. The voltage supplied to the bell is 120 volts. What is its resistance?

   R = ________ ohms.
2. A relay used to control a motor is rated at 25 ohms resistance. What voltage is required to operate the relay if it draws a current of 0.25 ampere?

   V = ________ volts.
3. An automobile battery supplies a current of 7.5 amperes to a headlamp with a resistance of 0.84 ohm. Find the voltage delivered by the battery.

   V = ________ volts.
4. What voltage is needed to light a lamp if the current required is 2 amperes and the resistance of the lamp is 50 ohms?

   V = ________ volts.
5. If the resistance of a radio receiver circuit is 240 ohms and it draws a current of 0.6 ampere, what voltage is needed?

   V = ________ volts.
6. A television circuit draws 0.15 ampere of current. The operating voltage is 120 volts. What is the resistance of the tv circuit?

   R = ________ ohms.
7. The resistance of the motor windings of an electric vacuum cleaner is 20 ohms. If the voltage is 120 volts, find the current drawn.

   I = ________ amperes.
8. The coil of a relay carries 0.05 ampere when operated from a 50-volt source. Find its resistance.

   R = ________ ohms.
9. How much current is drawn from a 12-volt battery when operating an automobile horn of 8 ohms resistance?

   I = ________ amperes.
10. Find the resistance of an automobile starting motor if it draws 90 amperes from the 12-volt battery.

    R = ________ ohms.
11. What current is drawn by a 5000-ohm electric clock when operated from a 120-volt line?

    I = ________ amperes.
12. Find the current drawn by a 50-ohm toaster from a 120-volt line.

    I = ________ amperes.

1-4

Basic Electrical Circuit Problems

Objective

Another fundamental understanding needed in any electrical field of study is that of basic electrical circuits. All electrical circuits are classified as either series, parallel, or combination (series-parallel) circuits. In this activity, you will review each of these basic electrical circuits and complete a self-test on each type of circuit.

Series Circuit Problems—Self-Test

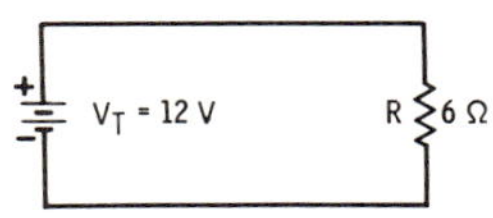

1. Find:

Current through R (I_R) = ______ amperes.

Voltage across R (V_R) = ______ volts.

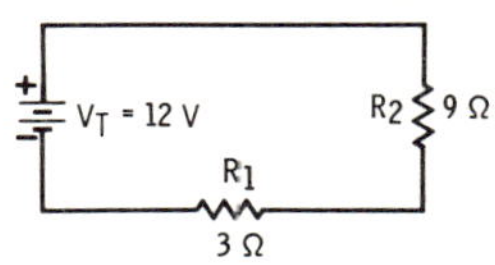

2. Find:

Total resistance (R_T) = ______ ohms.

Current through R_1 (I_{R_1}) = ______ amperes.

Current through R_2 (I_{R_2}) = ______ amperes.

Voltage across R_2 (V_{R_2}) = ______ volts.

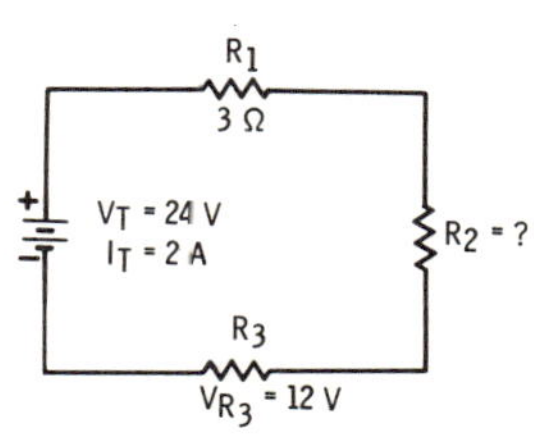

3. Find:

Voltage across R_1 (V_{R_1}) = ______ volts.

R_2 = ______ ohms.

Voltage across R_2 (V_{R_2}) = ______ volts.

R_3 = ______ ohms.

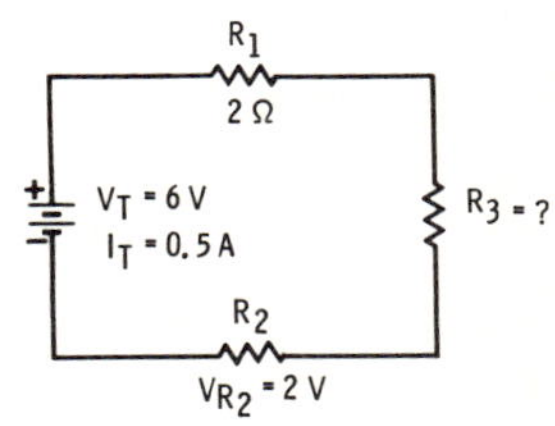

4. Find:

V_{R_1} = ______ volts. R_2 = ______ ohms.

V_{R_3} = ______ volts. R_3 = ______ ohms.

I_{R_3} = ______ amperes.

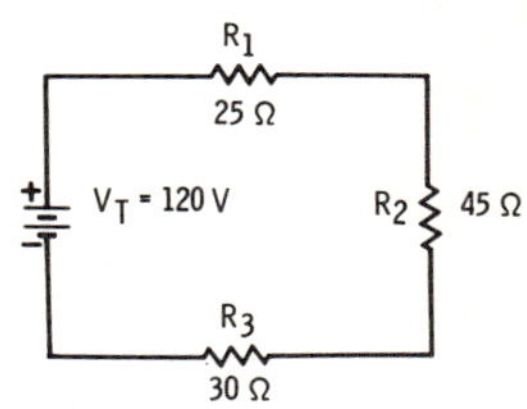

5. Find:

I_T = ______ amperes.

V_{R_1} = ______ volts.

V_{R_2} = ______ volts.

V_{R_3} = ______ volts.

Power converted by R_1

(P_{R_1}) = ______ watts.

Power converted by R_2

(P_{R_2}) = ______ watts.

Power converted by R_3

(P_{R_3}) = ______ watts.

Total power converted by circuit

(P_T) = ______ watts.

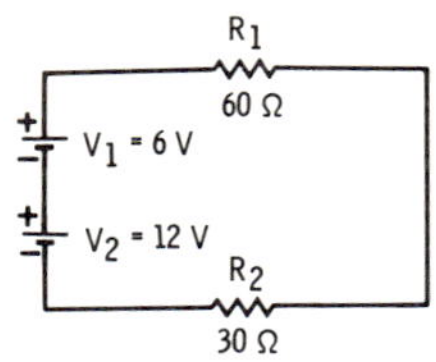

6. Find:

V_T = ______ volts. Total Power

(P_T) = ______ watts.

R_T = ______ ohms. V_{R_1} = ______ volts.

I_T = ______ amperes.

Parallel Circuit Problems—Self-Test

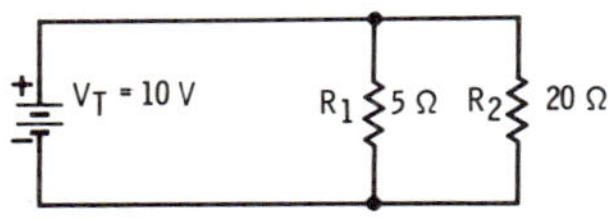

1. Find:

Total resistance (R_T) = ______ ohms.

Total current (I_T) = ______ amperes.

Current through resistor R_1

(I_{R_1}) = ______ amperes.

Current through resistor R_2

(I_{R_2}) = ______ amperes.

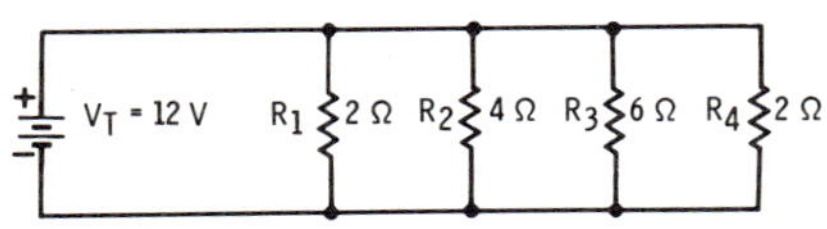

2. Find:

Total resistance (R_T) = ______ ohms.

Total current (I_T) = ______ amperes.

Current through resistor R_1

(I_{R_1}) = ______ amperes.

Current through resistor R_2

(I_{R_2}) = ______ amperes.

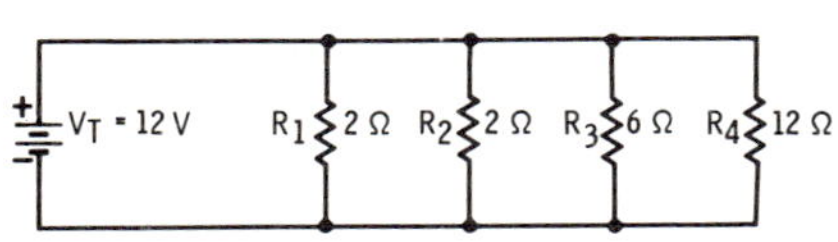

3. Find:

Total resistance (R_T) = ______ ohms.

Total current (I_T) = ______ amperes.

Current through resistor R_1

(I_{R_1}) = ______ amperes.

Current through resistor R_2

(I_{R_2}) = ______ amperes.

Current through resistor R_3

(I_{R_3}) = ______ amperes.

Current through resistor R_4

(I_{R_4}) = ______ amperes.

Total power (P_T) = ______ watts.

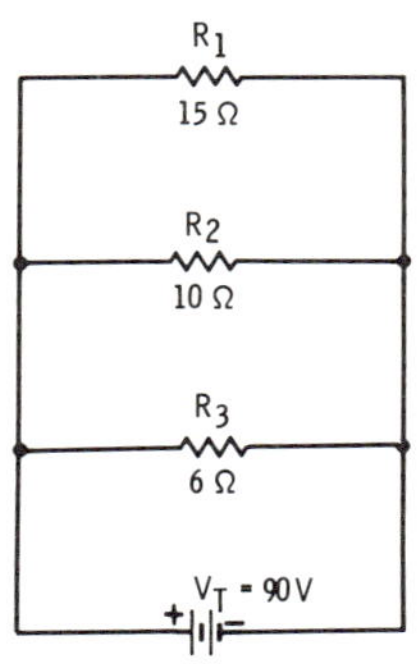

4. Find:

Total resistance (R_T) = ______ ohms.

Total current (I_T) = ______ amperes.

Current through resistor R_1

(I_{R_1}) = ______ amperes.

Current through resistor R_2

(I_{R_2}) = ______ amperes.

Current through resistor R_3

(I_{R_3}) = ______ amperes.

Power converted by resistor R_1

(P_{R_1}) = ______ watts.

Power converted by resistor R_2

(P_{R_2}) = ______ watts.

Power converted by resistor R_3

(P_{R_3}) = ______ watts.

5. Three resistors of 40 ohms, 90 ohms, and 75 ohms are connected in parallel across a 120-volt supply. Find:

   Total resistance (R_T) = ________ ohms.

   Total current (I_T) = ________ amperes.

   Current through each resistor:

   I_{R_1} = ________ amperes.

   I_{R_2} = ________ amperes.

   I_{R_3} = ________ amperes.

Combination (Series-Parallel) Circuit Problems—Self-Test

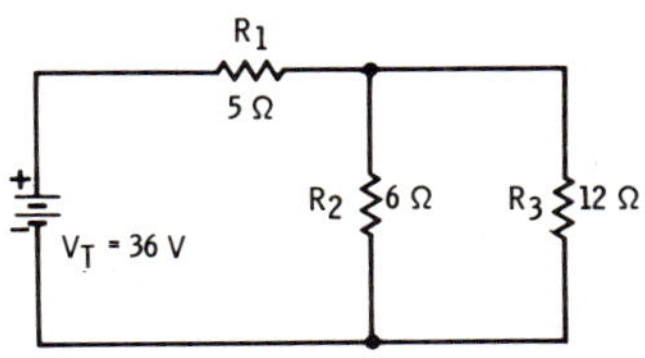

1. Find:

   Total resistance (R_T) = ________ ohms.

   Total current (I_T) = ________ amperes.

   Total power (P_T) = ________ watts.

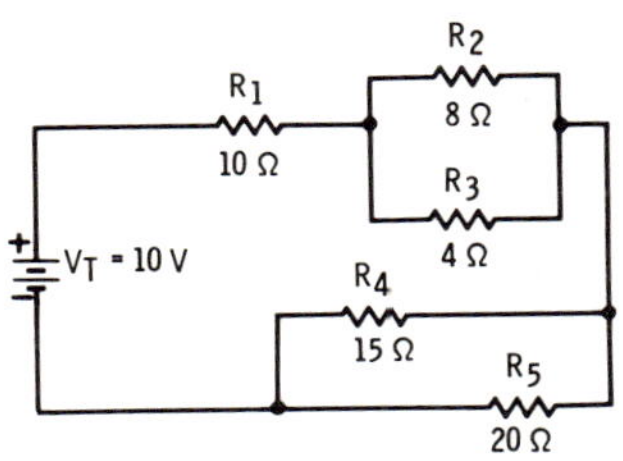

2. Find:

   Total resistance (R_T) = ________ ohms.

   Total current (I_T) = ________ amperes.

   Total power (P_T) = ________ watts.

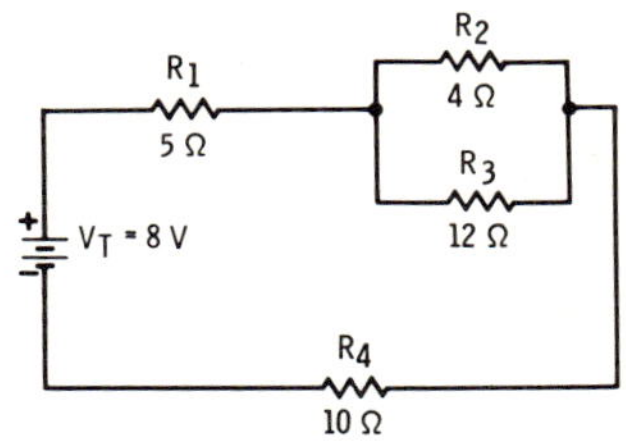

3. Find:

   Total resistance (R_T) = ________ ohms.

   Total current (I_T) = ________ amperes.

   Voltage across resistor R_1

   (V_{R_1}) = ________ volts.

   Voltage across resistor R_4

   (V_{R_4}) = ________ volts.

   Current through resistor R_2

   (I_{R_2}) = ________ amperes.

   Current through resistor R_3

   (I_{R_3}) = ________ amperes.

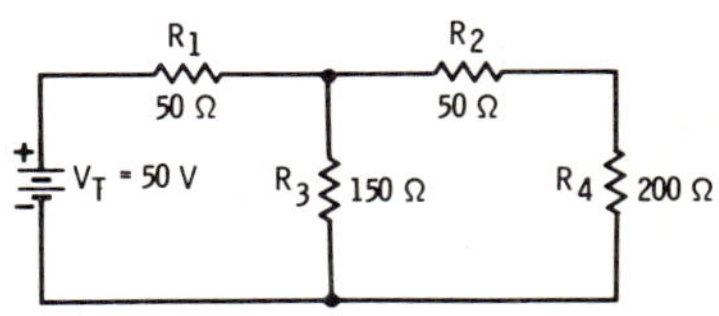

4. Find:

   Total resistance (R_T) = ________ ohms.

   Total current (I_T) = ________ amperes.

   Voltage across resistor R_1

   (V_{R_1}) = ________ volts.

1-5

Multifunction Meter Familiarization

Objective

In this activity, you will review the operational characteristics of the multifunction meter that you will use in the laboratory. Multifunction meters, such as the volt-ohm-milliammeter (vom), are used to measure voltage, resistance, and current. Thus, the term *multifunction* is used to describe the meter. Such meters are also multirange meters since several ranges are provided for various levels of each electrical quantity to be measured.

The multifunction meter will be used to make several types of measurements in the laboratory activities that follow. You should obtain an operation manual for the meter and review it before taking the self-test.

Multifunction Meter—Self-Test

Manufacturer of Meter: ______________

Model Number of Meter: ______________

Complete each of the following statements that pertain to the operation of the multifunction meter in your lab. If the statement is true or false, circle the T or F.

1. The multifunction meter is scaled to measure up to ____________ amperes of direct current.
2. Alternating current is read on the ____________ colored scales.
3. The OHMS ADJUST control should be checked each time the resistance range is changed. T F
4. To measure a direct current greater than 2 amperes, the selector is placed in the ____________ position.
5. For measuring current in a circuit, the meter should be connected in ____________.
6. For measuring voltage, the meter should be connected in ____________.
7. To measure 9 amperes of current, the selector must be placed in the ____________ position.
8. To measure 10 microamperes of current, the selector must be placed in the ____________ range.
9. To measure resistance, the red test lead must be placed in the jack marked ____________ and the black test lead in the jack marked ____________.
10. The most accurate resistance reading is found on the ____________ side of the meter scale.
11. Polarity is important when measuring ac voltage. T F
12. Polarity is not important when measuring resistance. T F
13. Up to ____________ volts dc can be measured with the multifunction meter.
14. For measuring a resistor valued at 10 ohms, the ____________ range would be used.
15. The top scale of the multifunction meter is used to measure ac voltage. T F
16. Polarities must be observed when measuring dc. T F
17. It is considered good practice to measure the resistance of a circuit with voltage applied to the circuit. T F
18. When measuring an unknown value of dc voltage, one should start at the highest scale and work down to the correct scale. T F
19. The meter used in the lab is a multifunction and also a multirange meter. T F
20. The multifunction meter has an overload circuit designed to protect the meter from damage. T F

Inductance Problems

Objective

Inductance is a very important circuit property. Many industrial systems, such as motors, transformers, and generators, have inductive characteristics. The basic construction of these systems causes them to have this property.

In this activity, you will solve some of the problems commonly associated with inductive circuits.

Inductance Problems

Solve the following problems by placing the correct answer in the blank spaces.

1. Total inductance in series: $L_1 = 2$ henrys, $L_2 = 3$ henrys, $L_3 = 2$ henrys.

   $L_T = L_1 + L_2 + L_3 =$ _______ henrys.

2. Total inductance in parallel: $L_1 = 2$ henrys, $L_2 = 3$ henrys, $L_3 = 8$ henrys.

   $$\frac{1}{L_T} = \frac{1}{L_1} + \frac{1}{L_2} + \frac{1}{L_3} = \_\_\_\_\_\_\_ \text{ henrys.}$$

3. Inductive reactance ($X_L = 2\pi fL$) for the following:

   (a) 30 Hz, 8 H. $X_L =$ _______ ohms.

   (b) 50 Hz, 2 H. $X_L =$ _______ ohms.

   (c) 60 Hz, 1.0 H. $X_L =$ _______ ohms.

   (d) 40 Hz, 3 H. $X_L =$ _______ ohms.

4. Mutual inductance in series to increase inductance (fields aiding):

   (a) $L_1 = 2$ henrys, $L_2 = 5$ henrys, $M = 0.55$.

   $L_T = L_1 + L_2 + 2M =$ _______ henrys.

   (b) $L_1 = 3$ henrys, $L_2 = 2$ henrys, $M = 0.35$.

   $L_T =$ _______ henrys.

5. Mutual inductance in series to decrease inductance (fields opposing):

   (a) $L_1 = 4$ henrys, $L_2 = 3$ henrys, $M = 0.85$.

   $L_T = L_1 + L_2 - 2M =$ _______ henrys.

   (b) $L_1 = 3$ henrys, $L_2 = 3$ henrys, $M = 0.4$.

   $L_T =$ _______ henrys.

6. Compute problems 4 and 5 if connected in parallel:

   (4a) $$\frac{1}{L_T} = \frac{1}{L_1 + M} + \frac{1}{L_2 + M}.$$

   $L_T =$ _______ henrys.

   (4b)

   $L_T =$ _______ henrys.

   (5a) $$\frac{1}{L_T} = \frac{1}{L_1 - M} + \frac{1}{L_2 - M}.$$

   $L_T =$ _______ henrys.

   (5b)

   $L_T =$ _______ henrys.

Capacitance Problems

Objective:

Capacitance is another very important circuit property. The electrostatic field associated with capacitive devices causes them to behave as they do. Capacitors are used for numerous industrial applications.

In this activity, you will solve some of the problems commonly associated with capacitive circuits.

Capacitance Problems

Solve the following problems by placing the answers in the blanks.

1. Compute the capacitance for the following capacitors, using the formula:

$$C = \frac{0.0885KA}{D},$$

where C is the capacitance in picofarads, K is the dielectric constant, A is the plate area in square centimeters, and D is the dielectric thickness in centimeters.

(a) Dielectric constant = 3.
(b) Size of plate = 2 × 200 centimeters.
(c) Thickness of dielectric = 0.03 centimeter.

C = ______ pF.

(a) Dielectric constant = 3.
(b) Size of plate = 3 × 200 centimeters.
(c) Thickness of dielectric = 0.04 centimeter.

C = ______ pF.

(a) Dielectric constant = 2.
(b) Size of plate = 0.25 × 275 centimeters.
(c) Thickness of dielectric = 0.02 centimeter.

C = ______ pF.

2. Compute the following for total capacitance in a series circuit, using the formula:

$$\frac{1}{C_T} = \frac{1}{C_1} + \frac{1}{C_2} + \frac{1}{C_3}.$$

(a) $C_1 = 10\ \mu F$
$C_2 = 10\ \mu F$
$C_3 = 30\ \mu F$

C_T = ______ μF.

(b) $C_1 = 40\ \mu F$
$C_2 = 20\ \mu F$
$C_3 = 20\ \mu F$

C_T = ______ μF.

(c) $C_1 = 80\ \mu F$
$C_2 = 60\ \mu F$
$C_3 = 80\ \mu F$

C_T = ______ μF.

3. Compute the capacitive reactance for an ac circuit of 60 Hz with a 30-μF capacitor, using the formula:

$$X_C = \frac{1}{2\pi fC} = \_\_\_\_\_\_ \text{ ohms.}$$

4. Compute the following for total capacitance in parallel, using the formula: $C_T = C_1 + C_2 + C_3$.

(a) $C_1 = 80\ \mu F$
$C_2 = 60\ \mu F$
$C_3 = 80\ \mu F$

C_T = ______ μF.

(b) $C_1 = 50\ \mu F$
$C_2 = 20\ \mu F$
$C_3 = 30\ \mu F$

C_T = ______ μF.

(c) $C_1 = 70\ \mu F$
$C_2 = 50\ \mu F$
$C_3 = 50\ \mu F$

C_T = ______ μF.

Basic Alternating-Current Values

Objective:

There are several basic ac values that are commonly used in electrical measurement and problem solving. These units are derived from ac sine wave relationships. The following ac sine wave values are often converted from one value to the other:

Effective value (rms) = 0.707 × peak value.
Average value = 0.636 × peak value.
Peak value = 1.41 × rms value.
Peak-to-peak value = 2.82 × rms value.

In this activity, you should review the preceding ac sine wave relationships and complete the self-test.

Basic AC Values—Self-Test

Solve the following problems by placing the correct answer in the blank space.

1. Compute the following effective values:
   Peak voltage:

   4 V ac = ______ V rms.

   12 V ac = ______ V rms.

   6 V ac = ______ V rms.

   10 V ac = ______ V rms.

   1.5 V ac = ______ V rms.

   2 V ac = ______ V rms.

   5 V ac = ______ V rms.

   9 V ac = ______ V rms.

2. Compute the following peak and peak-to-peak (p-p) values:
   Rms voltage:

   3 V ac = ______ V peak; ______ V p-p.

   8 V ac = ______ V peak; ______ V p-p.

   7 V ac = ______ V peak; ______ V p-p.

   9 V ac = ______ V peak; ______ V p-p.

   10 V ac = ______ V peak; ______ V p-p.

   15 V ac = ______ V peak; ______ V p-p.

   11 V ac = ______ V peak; ______ V p-p.

   18 V ac = ______ V peak; ______ V p-p.

Alternating-Current Circuit Calculations

Objective

For more practice in solving ac circuit problems, you should complete this activity. It is important to have an understanding of basic ac circuit problems in order to solve more advanced circuit problems. These problems may be referred to when a similar problem is encountered in later activities. The organization of the activity is such that one example of each type of ac circuit calculation is done.

AC Circuit Calculations

1. Series inductance: $L_T = L_1 + L_2 + L_3$.

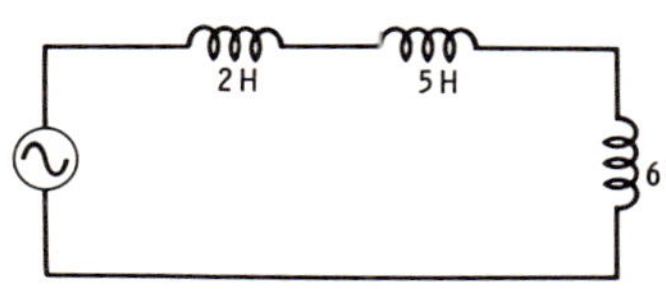

What is the inductance of this circuit?

L = ______ henrys.

2. Parallel inductance: $\frac{1}{L_T} = \frac{1}{L_1} + \frac{1}{L_2} + \frac{1}{L_3}$.

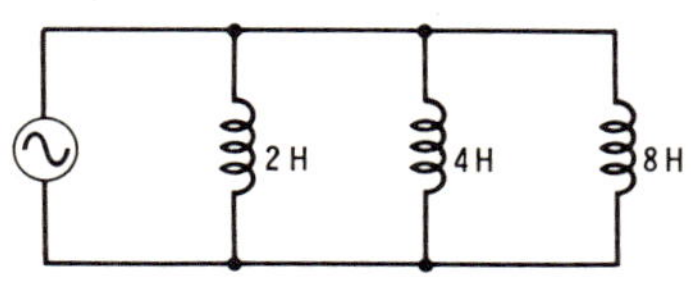

What is the inductance of this circuit?

L = ______ henrys.

3. Inductive reactance: $X_L = 2\pi fL$.

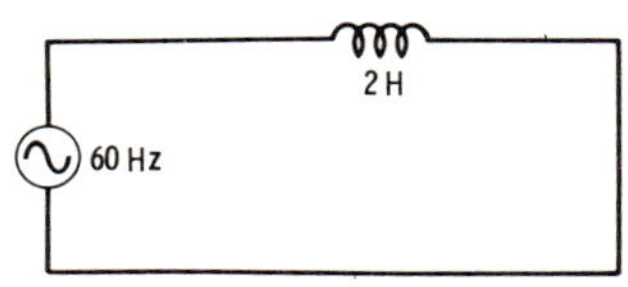

What is the inductive reactance?

X_L = ______ ohms.

4. Series capacitance: $\frac{1}{C_T} = \frac{1}{C_1} + \frac{1}{C_2} + \frac{1}{C_3}$.

F = farads
μF = microfarads (10^{-6})
pF = picofarads (10^{-12})

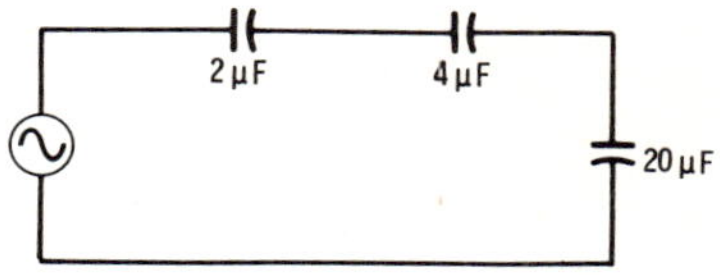

What is the capacitance?

C = ______ farads.

5. Parallel capacitance: $C_T = C_1 + C_2 + C_3$.

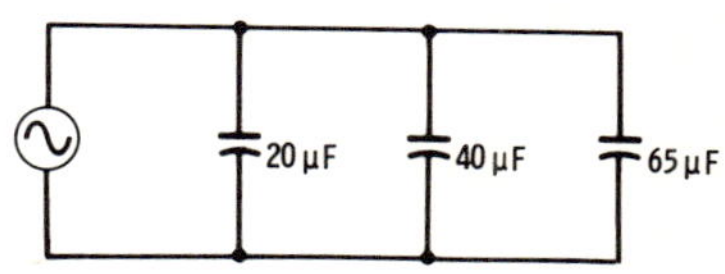

What is the capacitance?

C = ______ μF.

6. Capacitive reactance: $X_C = \frac{1}{2\pi fC}$.

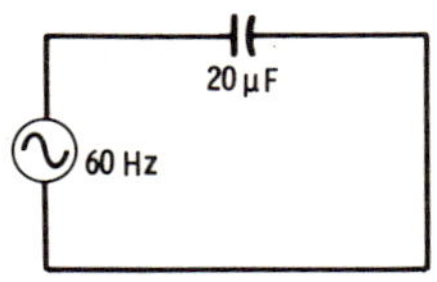

What is the capacitive reactance?

X_C = ______ ohms.

7. RL time constant: t (seconds) = L/R.

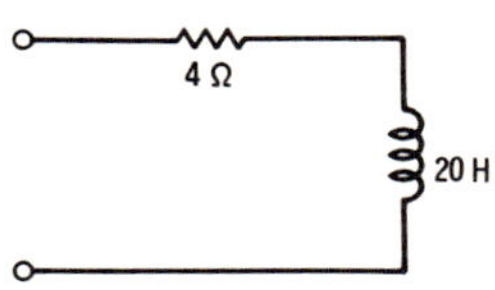

What is the time constant?

t = ______ seconds.

8. RC time constant: t = RC.

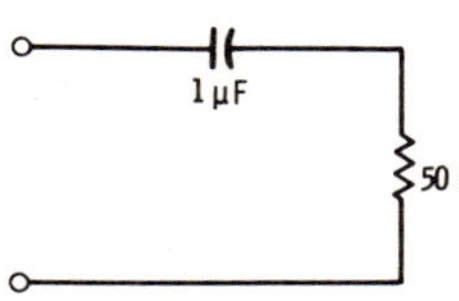

What is the time constant?

t = ______ seconds.

9. Impedance in RC or RL circuit: $Z = \sqrt{R^2 + (X_C)^2}$ or $\sqrt{R^2 + (X_L)^2}$.

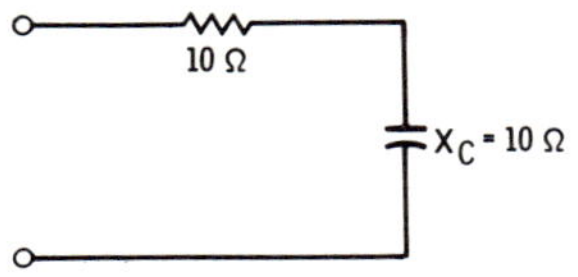

What is the impedance?

Z = ______ ohms.

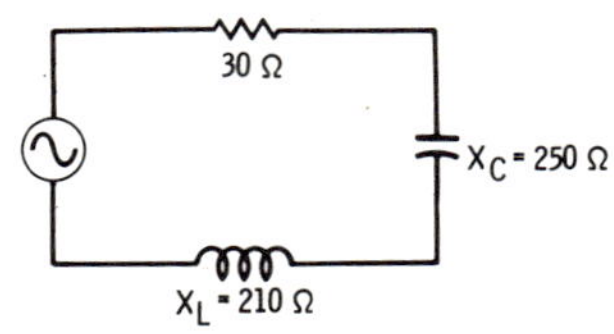

What is the impedance?

Z = ______ ohms.

10. Impedance in RLC Circuits:
$Z = \sqrt{R^2 + (X_L - X_C)^2}$

1-10

Electronic Multifunction Meter

Objective

Electronic multifunction meters are used to make measurements in circuits where the "loading" effect of a meter may cause erroneous readings. The input impedance of an electronic multifunction meter, such as a digital voltmeter (dvm), transistorized voltmeter (tvm), or vacuum-tube voltmeter (vtvm), is ordinarily much higher than that of a vom or similar nonelectronic meter. The disadvantage of this type of meter is that an ac voltage source is usually required; therefore, it is not as portable as the vom.

In this activity, you will review the operational characteristics of the electronic multifunction meter that you will use in the lab. You should obtain an operation manual for the meter and review it before taking the self-test.

Electronic Multifunction Meter—Self-Test

Complete each of the following statements that pertain to the operation of the electronic multifunction meter in your lab. If the statement is true or false, circle the T, F, or NA (not applicable).

1. Resistance is read on the top scale of the meter. T F NA
2. The red colored scales are used to measure ac voltage. T F NA
3. Current, either ac or dc, can be measured with the meter. T F
4. There are several quantities that can be measured with the meter. They are: ________________________________.
5. The probe of the vtvm is considered negative when measuring dc. T F
6. Peak-to-peak voltages may be read directly on the meter. T F
7. When measuring resistance, all power in the circuit must be turned off. T F
8. The meter can be "zero-centered." T F
9. The meter can measure both positive and negative dc voltage. T F
10. When measuring resistance, the pointer deflects from left to right. T F NA
11. When measuring any value, one should start with the highest range and work down to the proper scale. T F
12. The OHMS ADJUST control must be readjusted each time a new range is selected for measuring resistance. T F
13. When measuring dc, polarity must be observed. T F
14. Polarity is not important when measuring ac. T F
15. The meter has an overload circuit designed to protect the meter. T F

1-11

Oscilloscope Familiarization

Objective

In this activity, you will review the operational characteristics of the oscilloscope that you will be using in the lab. Since there are so many types of oscilloscopes in use today, the self-test will be very general. You should keep in mind that some controls listed on this self-test may be different from those on your oscilloscope.

The oscilloscope will be used to observe many types of alternating-current waveforms in the following lab activities. You should look at the operation manual for the scope you are using, if it is available, before taking the self-test.

Oscilloscope Controls—Self-Test

Manufacturer of Oscilloscope: ______________________________

Model Number of Oscilloscope: ______________________________

Section A

In the following, match the oscilloscope control with the proper function. If the control on your scope has a different name, place it in parentheses beside the control name listed.

_____ 1. INTENSITY

_____ 2. FOCUS

_____ 3. VERTICAL POSITION

_____ 4. HORIZONTAL POSITION

_____ 5. VERTICAL GAIN

_____ 6. HORIZONTAL GAIN

_____ 7. VERTICAL ATTENUATION

_____ 8. SWEEP SELECT

_____ 9. VERNIER

_____ 10. SYNCHRONIZATION SELECT

_____ 11. HORIZONTAL INPUT

_____ 12. VERTICAL INPUT

A. Selects horizontal sweep frequency.
B. Provides up and down adjustment of beam.
C. Determines amplitude of horizontal deflection.
D. Controls size of electron beam.
E. Provides left and right adjustment of beam.
F. Controls brightness of beam.
G. Eliminates vertical shift of display.
H. External voltages applied to this control.
I. Determines frequency of the sawtooth sweep.
J. Determines amplitude of vertical deflection.
K. Setting for calibration voltage.
L. Controls external signals to vertical amplifier.
M. Determines amount of voltage used to synchronize sweep oscillator.
N. Selects type of synchronization desired.
O. Balances dc output.
P. Sweeps at line frequency.

Section B

Define the following terms that are related to oscilloscope use with alternating-current signals applied.

1. cycle:

2. frequency:

3. hertz:

4. rms voltage:

5. peak-to-peak voltage:

6. kilohertz:

7. megahertz:

8. amplitude:

Trigonometry Problems

Objective

In this activity, you will use the trigonometry table to find the sine, cosine, and tangent values. You will also find the values of angles when a sine, cosine, or tangent value is given. A knowledge of trigonometry is particularly important in dealing with ac circuits.

Trigonometry Problems

Find the values of the following trig functions by referring to a "table of trig ratios."

1. sin 138° = __________
2. sin 62° = __________
3. sin 212° = __________
4. cos 152° = __________
5. tan 141° = __________
6. sin 222° = __________
7. cos 230° = __________
8. tan 201° = __________
9. sin 318° = __________
10. cos 290° = __________
11. tan 310° = __________
12. arc sin 0.4226 = __________
13. arc cos 0.4848 = __________
14. arc tan 9.5144 = __________
15. arc tan 0.2309 = __________
16. arc sin 0.9613 = __________
17. arc cos 0.2924 = __________
18. arc sin −0.8660 = __________
19. arc cos −0.6820 = __________
20. arc tan −3.7321 = __________

Abbreviations:

sin = sine
cos = cosine
tan = tangent
arc sin = angle whose sine is
arc cos = angle whose cosine is
arc tan = angle whose tangent is

2-1

National Electrical Code Familiarization

Objective

The *National Electrical Code* (NEC) is a very important document to understand. All industrial equipment and wiring must conform to the NEC standards. The NEC is not difficult to use. The user should become familiar with the comprehensive index contained in the NEC and the organization of the various sections. For instance, if you wished to review the standards related to "system grounding," you should look in the index and locate this term. The index will refer you to the appropriate sections in the NEC which discuss "system grounding."

In this activity, you will need to review the latest edition of the NEC. After you have learned how to use it, you should complete the self-test on the NEC.

National Electrical Code—Self-Test

Using your copy of the NEC, find the correct responses to complete each of the following statements:

1. A new edition of the NEC is published every ______________ years.
2. A No. 8 copper conductor is capable of conducting ______________ amperes of current.
3. When selecting nonmetallic-sheathed cable for general use, you should specify Type ____________.
4. If you were wiring for an air conditioner, you should refer to Article ______________ in the NEC.
5. The diameter of electrical conductors is measured with a ______________ gauge.
6. Line-to-ground shock hazards may be reduced by utilizing __________________ __________________ __________________.
7. The definition of a "ground" may be located in Article ______________ of the NEC.
8. Article 402 of the NEC specifies that Type CF fixture wire has ______________ insulation.
9. If you were installing a 3-phase motor with associated control circuits, you should refer to Article ______________ of the NEC.
10. Article 810 of the NEC provides electrical wiring specifications for ______________.
11. Underground feeder cable is available in sizes ______________ to ______________ inclusive.
12. The full-load current of a 1-horsepower, 120-volt dc motor as listed in the NEC is ______________ amperes.
13. The full-load current of a 30-horsepower, 3-phase ac induction motor operating on 208 volts is ______________ amperes. (See information at bottom of table for conversion.)
14. According to Article 210 of the NEC, grounded conductors in a branch circuit should be color coded ______________ or ______________.
15. The general lighting load of an office building is specified to be ______________ watts per square foot of floor space.

Occupational Safety and Health Act (OSHA) Familiarization

Objective

The Occupational Safety and Health Act (OSHA) has had a profound impact on industrial operation. Due to this impact, it is necessary for all industrial employees to be familiar with some of the OSHA standards. In order to familiarize you with OSHA industrial standards, you will be able to complete an OSHA inspection by using the checklist that follows. You may use this checklist for your laboratory or any shop facility. A safe and healthy environment is a necessity for industrial operation.

Inspection Checklist—Student Activity

Inspections that involve checks similar to those listed here should be performed periodically in industry. You may perform this brief inspection individually or in groups. This is *not* a detailed inspection.

For the items on the checklist, place the number to the left of the item that describes its condition, as follows:

0 = Not applicable.
1 = Very good (needs no attention).
2 = Good (needs little attention).
3 = Acceptable (needs some attention).
4 = Bad (needs attention soon).
5 = Very bad (needs immediate attention).

Place necessary comments to the right of each item.

Section A. General Plant Environment

_____ 1. Temperature.

_____ 2. Ventilation.

_____ 3. Illumination.

_____ 4. Condition of equipment.

_____ 5. Condition of stairs.

_____ 6. Condition of floors.

_____ 7. Condition of walls.

_____ 8. Condition of windows.

_____ 9. Condition of ceilings.

_____ 10. Availability of fire extinguishers.

_____ 11. Condition of fire extinguishers.

_____ 12. Fire escape plan.

_____ 13. Hanging objects on walls or ceiling.

_____ 14. Electrical power receptacles.

_____ 15. General appearance (neatness).

_____ 16. Condition of benches.

_____ 17. Condition of supply room.

_____ 18. Storage of materials.

_____ 19. Other ______________________.

Section B. Equipment

_____ 1. High voltage and danger areas identified.

_____ 2. Machines and equipment well-guarded and protected.

_____ 3. All moving gears, belts, etc. are enclosed properly.

_____ 4. Storage area for tools and materials used with equipment.

_____ 5. Tools in good condition.

_____ 6. Machines in good condition for operation.

_____ 7. Good access to equipment control devices.

_____ 8. Access to circuit breaker to turn off all equipment.

_____ 9. Other ______________________.

Section C. Electrical Distribution

_____ 1. No temporary wiring used.

_____ 2. Switches close to equipment they control.

_____ 3. Switches all enclosed properly.

_____ 4. Power outlets well identified.

_____ 5. Equipment wired according to NEC specifications.

_____ 6. Necessary overload protection on equipment.

_____ 7. Availability of master switch for all equipment.

_____ 8. Other ______________________.

Section D. Miscellaneous

_____ 1. Availability of first-aid kit.

_____ 2. Availability of accident reports.

_____ 3. Knowledge of accident reporting procedure.

_____ 4. Evidence of safety awareness.

_____ 5. Safety glasses and protective equipment available where needed.

_____ 6. Waste materials disposed of easily.

_____ 7. Other ______________________.

Section E. General Safety Rating of Facility: _______

(Comments):

Industrial Standards

Objective

The daily operation and the manufactured products of industry must meet certain standards. These standards are prescribed by such organizations as the American National Standards Institute (ANSI) and the Institute of Electrical and Electronic Engineers (IEEE).

In this activity, you will examine one of the ANSI standards booklets. Much information is contained in the ANSI standards. You should become familiar with the index for standards. The following procedure will familiarize you with the ANSI standards for "Electrical and Electronics Diagrams."

Procedure

1. Locate the ANSI standard for "Electrical and Electronics Diagrams."
2. What is the number of this standard?

____

3. When was it published?
4. By whom was it published?
5. From the "Table of Contents," list five major topics of this standard:

____

____

____

____

____

6. How does the ANSI standard define a "schematic diagram"?

____

____

____

7. How is a "one-line diagram" defined?

____

____

____

8. What is a "continuous-line" connection diagram?

____

____

____

9. What is a "highway or cable diagram"?

____

____

____

10. How is a "terminal diagram" defined?

____

____

____

2-4

Scientific Notation and Unit Conversion

Objective

In the study of industrial electrical systems, it is often necessary to convert numbers from one unit to another in order to solve problems. In this activity, you will study scientific notation and unit conversion. You will then be able to successfully complete a self-test.

Scientific Notation

The technique of using powers of 10 can greatly simplify mathematical calculations. A number containing many zeros to the right or to the left of the decimal point can be dealt with much more easily when put in the form of powers of 10. For example, $0.0000027 \times 0.000028$ can be handled more easily when put in the form $2.7 \times 10^{-6} \times 2.8 \times 10^{-5}$. Notice the number of places that the decimal point is moved in each situation.

Table 2-4A lists some of the powers of 10. In a whole number, the power to which the number is raised is positive and equals the number of zeros following the one. In decimals, the power is negative and equals the number of places the decimal point is moved to the left.

Table 2-4A. Powers of 10

Number	Power of 10
1,000,000	10^6
100,000	10^5
10,000	10^4
1000	10^3
100	10^2
10	10^1
1.0	10^0
0.1	10^{-1}
0.01	10^{-2}
0.001	10^{-3}
0.0001	10^{-4}
0.00001	10^{-5}
0.000001	10^{-6}

Any number written as a multiple of a power of 10 and a number between one and ten is said to be expressed in scientific notation. For example:

$$91{,}000{,}000 = 9.1 \times 10{,}000{,}000 \text{ or } 9.1 \times 10^7$$
$$800{,}000{,}000 = 8 \times 100{,}000{,}000 \text{ or } 8 \times 10^8$$
$$0.0000000007 = 7 \times 0.0000000001 \text{ or } 7 \times 10^{-10}$$

Scientific notation greatly simplifies the multiplication and division of large numbers or small decimals. For example:

Multiply $1800 \times 0.000015 \times 300 \times 0.0048$:

$$= 1.8 \times 10^3 \times 1.5 \times 10^{-5} \times 3 \times 10^2 \times 4.8 \times 10^{-3}$$
$$= 1.8 \times 1.5 \times 3 \times 4.8 \times 10^{3-5+2-3}$$
$$= 38.88 \times 10^{-3}$$
$$= 0.03888$$

Divide 75,000 by 0.0005:

$$= \frac{7.5 \times 10^4}{5 \times 10^{-4}}$$
$$= \frac{7.5}{5} \times 10^{4-(-4)}$$
$$= 1.5 \times 10^8$$
$$= 150{,}000{,}000$$

A number will often have to be converted from one unit to another. This is true particularly when applying Ohm's law for solutions of basic problems. For example, if you know that the voltage applied to a circuit is 10 volts and the current is 10 milliamperes, to apply the Ohm's law relationship, you must convert 10 milliamperes to 0.01 ampere by moving the decimal point three places to the left. Unit conversion may be accomplished by using the conversion chart contained in Chapter 2 of *Industrial Electrical Systems.*

Scientific Notation and Unit Conversion—Self-Test

Convert each of the following listed quantities to scientific notation, then to the unit indicated. Place the correct responses in the appropriate blank spaces.

	Scientific Notation	*Conversion*
1. 0.025 ampere to milliamperes	2.5×10^{-2}	25 mA
2. 250 microamperes to amperes	________	________
3. 15,000 ohms to megohms	________	________

4. 0.65 megohm to ohms ____________ ________
5. 370 microvolts to millivolts ____________ ________
6. 6000 watts to kilowatts ____________ ________
7. 3,520,000 ohms to megohms ____________ ________
8. 5.24 volts to millivolts ____________ ________
9. 5129 microamperes to amperes ____________ ________
10. 3.21 milliamperes to microamperes ____________ ________
11. 500 amperes to microamperes ____________ ________
12. 0.005 microampere to milliamperes ____________ ________
13. 6225 volts to kilovolts ____________ ________
14. 122 ohms to kilohms ____________ ________
15. 18,000,000 ohms to megohms ____________ ________
16. 358 microvolts to volts ____________ ________
17. 5820 milliamperes to amperes ____________ ________
18. 22 microamperes to amperes ____________ ________
19. 8250 watts to megawatts ____________ ________
20. 6.28 megohms to ohms ____________ ________

Metric Conversions

Objective

The importance of the metric system of measurement must be stressed in industry. Conversion of units from the U.S. system to the metric system is easy to perform. In this activity, you will become familiar with the metric system of measurement and complete a self-test dealing with metric conversions.

Metric Conversion—Self-Test

1. What are the basic units of the metric system?

____

____

2. Complete the following conversions:

(a) 1 meter = ______ centimeters.

(b) 1 centimeter = ______ millimeters.

(c) 5000 grams = ______ kilograms.

(d) 2 liters = ______ milliliters.

(e) 1 meter = ______ inches.

(f) 1 centimeter = ______ inches.

(g) 1 mile = ______ kilometers.

(h) 1 gram = ______ ounces.

(i) 30 ounces = ______ grams.

(j) 1 kilogram = ______ pounds.

(k) 3 liters = ______ quarts.

(l) 1 gallon = ______ liters.

(m) 10 cups = ______ liters.

(n) 50 miles = ______ kilometers.

(o) 2 cubic centimeters (cm^3) = ______ cubic inches (in^3).

(p) 4 cubic feet (ft^3) = ______ cubic meters (m^3).

2-6

Industrial Symbols and Diagrams

Objective

In Activity 1-2, basic electrical symbols were reviewed. There are several symbols that are unique to industrial applications. In this activity, you will examine some common industrial symbols and diagrams. You will show your knowledge of symbols and diagrams by taking a self-test. You should refer to the symbols and diagrams in the Appendix of *Industrial Electrical Systems* for your review study.

Industrial Symbols—Self-Test

In the space next to each listed device, draw the proper symbols for that device.

1. Incandescent lamp
2. Mercury vapor lamp
3. Fluorescent lamp
4. Circuit breaker
5. Fuse
6. Push-button switch (normally open)
7. Push-button switch (normally closed)
8. Pressure-actuated switch
9. Temperature-actuated switch
10. Thermal cutout
11. Normally closed contact
12. Normally open contact

Industrial Diagrams—Self-Test

The most common types of diagrams used to convey the necessary information about the components, circuits, and subsystems of an operational industrial system are: (1) connection diagrams, (2) block diagrams, (3) schematic diagrams, and (4) one-line diagrams.

In this activity, you will need to do some research to locate some simple industrial systems that may be represented by using the appropriate type of diagram.

1. *Connection Diagram:* The connection diagram primarily shows the physical relationships of the parts of a system. It is used to identify various parts for installation or servicing. The wires that interconnect circuits together are ordinarily shown in their correct location. In the following space, draw a simple connection diagram for some industrial system:

2. *Block Diagram:* The block diagram is a very general method used to show the way subsystems of any industrial system fit together to form a functional unit. Rectangles are used to represent subsystems and arrows are used to connect the blocks and show the interrelationships of subsystems. In the following space, draw a block diagram of some industrial system:

3. *Schematic Diagram:* The schematic diagram is used to show the electrical circuit relationships of the various components that comprise an operational system. The electrical, but not necessarily the physical, layout of the circuit is shown. This type of industrial diagram is the most often used. In the laboratory activities that follow, this type of diagram will be used predominantly. In the following space, draw a schematic diagram of some industrial circuit:

4. *One-Line Diagram:* The one-line diagram is a simplified form of the schematic diagram. This type of diagram is used to symbolically show how the components of a system fit together. It is similar to the block diagram in several respects; however, actual symbols, rather than rectangular shapes are used to show component parts. In the following space, draw a one-line diagram of some industrial system:

3-1

Single-Phase AC Generator (Alternator) Analysis

Objective

In this laboratory activity, you will observe the construction and operation of a single-phase alternator. This type of power source is used to produce alternating current for various applications.

The generation of alternating current is influenced by armature speed, field strength, and number of coil turns. An additional consideration in electric generators, which may be just as important as the above three prime factors, is the load. The load is referred to as the external resistance or impedance that is connected across the generator output. In this activity, a lamp will be used as the load. The effects of load changes on generator output will be studied.

In the study of electrical machinery, the term *armature* is used frequently. This term needs mentioning as a point of information for this experiment. An armature is defined as the part of a rotating electrical machine that has voltage induced or generated into it. Most alternators have a stationary armature and a rotating field.

In this activity, you may use either a commercially manufactured, single-phase ac generator or one that has been constructed in the lab.

Equipment

Single-phase ac generator.
Multifunction meter.
Prime mover.
Dc power source.
Lamps—40 and 60 W.
Ac current meter.
Capacitor—2 μF.
Inductor—15 H.

Procedure

1. A connection diagram for a single-phase ac generator is shown in Fig. 3-1A.

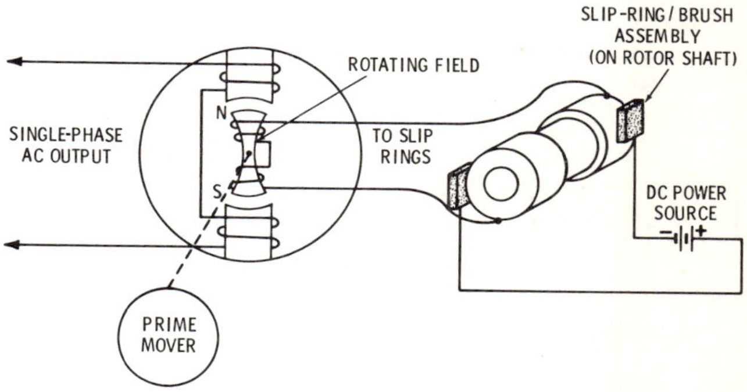

Fig. 3-1A. Connection diagram for a single-phase ac generator.

2. Obtain a single-phase rotating field alternator. Use a variable speed prime mover.
3. In this part of the activity, alternator output values will be measured as the speed is varied.
   A. Connect a 40-watt lamp across the alternator output. Measure the resistance of the lamp.

      R = ________ ohms.
   B. Apply approximately 12 volts dc to the rotating field.
   C. Vary the speed of the prime mover in 250-r/min increments and complete the measurements indicated in Table 3-1A. Calculate the output power as V^2/R.

Table 3-1A. Alternator Output Versus Speed (40-W Load)

Speed (r/min)	Output Voltage	Output Current	Output Power
0			
250			
500			
750			
1000			

4. In this part of the activity, alternator output values with a 60-watt lamp load connected will be compared to the values obtained in the previous section.
   A. Replace the 40-watt lamp load with a 60-watt lamp load. Measure the resistance of this lamp.

      R = ______ ohms.
   B. Maintain the same dc excitation as in Step 3 of the experiment.
   C. Complete the measurements and calculations indicated in Table 3-1B.

Table 3-1B. Alternator Output Versus Speed (60-W Load)

Speed (r/min)	Output Voltage	Output Current	Output Power
0			
250			
500			
750			
1000			

5. In this part of the activity, alternator output values will be measured as the dc excitation voltage is varied.
   A. Connect a 40-watt lamp load across the output.
   B. Maintain a prime-mover speed of about 500 r/min.
   C. Vary the dc excitation voltage in 4-volt increments and complete the measurements and calculations indicated in Table 3-1C.

Table 3-1C. Alternator Output Versus DC Excitation Voltage

Excitation Voltage	Output Voltage	Output Current	Output Power
0			
4			
8			
12			
16			
20			

6. In this part of the activity, alternator output values will be measured with an inductive load added across the output.
   A. Connect a 15-henry inductor in series with the 40-watt lamp load as shown in Fig. 3-1B.
   B. Use the same dc excitation values as in Step 3.
   C. Complete Table 3-1D. Calculate true power as V_R^2/R.

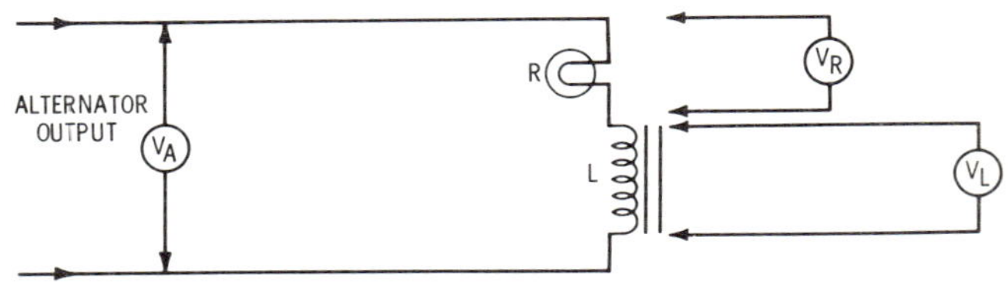

Fig. 3-1B. Inductive load in series with resistive 40-watt lamp load.

7. In this part of the activity, alternator output will be measured with a capacitive load connected in series with the resistive and inductive loads of the previous section. The addition of a capacitive load is referred to as *power factor correction.*
   A. Connect a 2-μF capacitor in series with the lamp and inductor used in the previous section (see Fig. 3-1C).
   B. Keep the dc excitation the same value.
   C. Complete Table 3-1E.

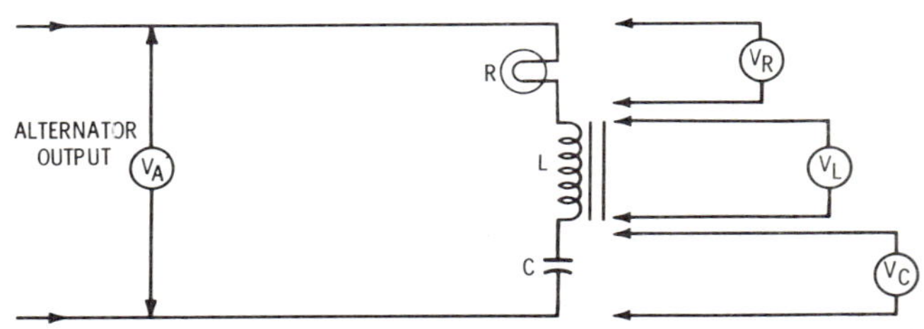

Fig. 3-1C. Capacitive load in series with resistive and inductive loads.

Analysis

1. What are the types and characteristics of single-phase ac generators?

____

____

2. What is meant by power factor correction?

____

____

3. Compare Table 3-1A to Table 3-1B in terms of alternator output when the load is changed from 40 to 60 watts.

____

____

Table 3-1D. Alternator Output Versus Speed (RL Load)

Speed (r/min)	Voltage Across Lamp (V_R)	Voltage Across Inductor (V_L)	Voltage Across Alternator Output (V_A)	True Power
0				
250				
500				
750				
1000				

4. From the data of Table 3-1C, how does dc excitation affect alternator output?

____

____

5. From Table 3-1D, with speed = 500 r/min, draw a vector diagram of the voltages in the circuit.

6. From Table 3-1E, with speed = 500 r/min, draw a vector diagram of the voltages in the circuit.

7. Compare the true power values of Tables 3-1D and E. Why do they differ?

____

____

Table 3-1E. Alternator Output Versus Speed (RLC Load)

Speed (r/min)	V_R	V_L	V_C	V_A	True Power
0					
250					
500					
750					
1000					

3-2

Three-Phase AC Generator (Alternator) Analysis

Objective

In this laboratory activity, you will observe the operational characteristics of a three-phase ac alternator. This type of alternator is used to supply most of the ac power used today. Power plants throughout the country use three-phase alternators. In fact, automobiles use this type of alternator in their charging systems. The three-phase output must be rectified to produce dc for charging the automobile battery. You may use a commercially manufactured three-phase alternator, or one that has been constructed in the lab for the analysis that follows.

Equipment

Three-phase ac generator.
Prime mover.
Lamps—40 and 60 W (3 each).
Multifunction meter.
Ac current meter.
Tachometer.
Inductors—8 H (3).
Capacitors—1.0 μF (3).
Variable dc power source.

Procedure

1. A connection diagram for a three-phase ac generator is shown in Fig. 3-2A.
2. Obtain a three-phase alternator and connect its shaft to a variable-speed prime mover to produce rotation. First, connect the stator coils in a "delta" configuration.

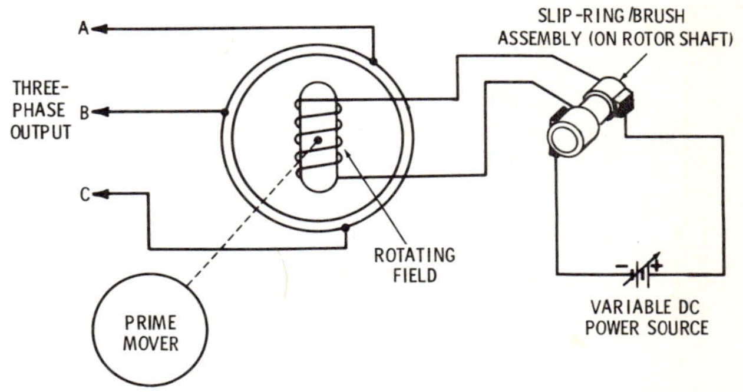

Fig. 3-2A. Connection diagram for a three-phase ac generator.

3. In this part of the experiment, alternator output values will be measured as the speed is varied.
   A. Connect the 40-watt lamps in a delta configuration across the alternator output. Measure the resistance of one of the lamps.

      R = ______ ohms.

   B. Apply approximately 12 volts dc to the rotating field.
   C. Vary the speed of the prime mover in 250 r/min increments and complete Table 3-2A. Calculate the output power per phase as V_p^2/R and the total power by adding the P_p values.
   D. Perform the same procedure for 60-watt lamp loads in a delta configuration and complete Table 3-2B.
4. In this part of the experiment, alternator output values with a "wye"-connected stator and load will be compared to the values obtained in the previous part of the experiment.

Table 3-2A. Delta Connection (40-W Load)

Speed (r/min)	Phase Voltages (V_p)			Phase Currents (V_p/R)			Power per Phase			Total Power
	A	B	C	A	B	C	A	B	C	
250										
500										
750										
1000										

Table 3-2B. Delta Connection (60-W Load)

Speed (r/min)	Phase Voltages (V_p)			Phase Currents (V_p/R)			Power per Phase			Total Power
	A	B	C	A	B	C	A	B	C	
250										
500										
750										
1000										

A. Rewire the alternator stator and load in a wye configuration.
B. Maintain the same dc excitation as in Step 3 of the experiment.
C. Complete Table 3-2C.
D. Perform the same procedure for 60-watt lamp loads and complete Table 3-2D.

5. In this part of the experiment, alternator output values will be measured as the dc excitation voltage is varied. Keep the assembly the same as in Step 4.
   A. Connect 40-watt lamp loads across the output.
   B. Maintain a prime mover speed of about 500 r/min.
   C. Vary the dc excitation voltage in 4-volt increments and complete Table 3-2E.
6. In this part of the experiment, alternator output values will be measured with an inductive load added across the output as shown in Fig. 3-2B. Use the wye connection.
   A. Connect 15-henry inductors in parallel with the 40-watt lamp loads of each phase.
   B. Use the same dc excitation values as in Steps 3 and 4.
   C. Complete Table 3-2F. Calculate total power per phase as V_p^2/R. Use only one phase for measurements as shown in Fig. 3-2B. Measure

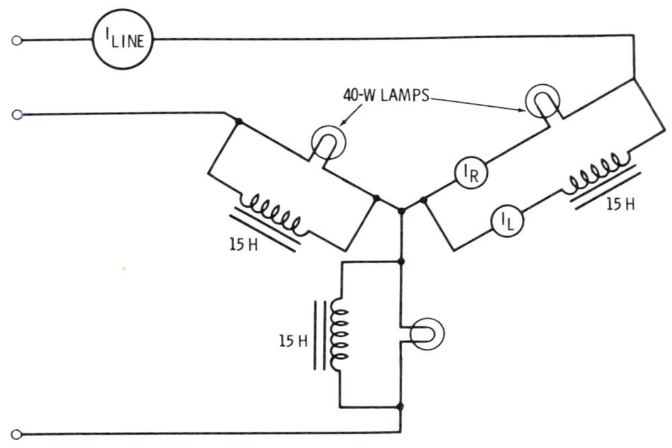

Fig. 3-2B. Inductive loads in parallel with resistive loads.

Table 3-2C. Wye Connection (40-W Load)

Speed (r/min)	Phase Voltages (V_p)			Line Voltages (V_L)			Phase Currents (V_p/R)			Power per Phase			Total Power
	A	B	C	A	B	C	A	B	C	A	B	C	
250													
500													
750													
1000													

Table 3-2D. Wye Connection (60-W Load)

Speed (r/min)	Phase Voltages (V_p)			Line Voltages (V_L)			Phase Currents (V_p/R)			Power per Phase			Total Power
	A	B	C	A	B	C	A	B	C	A	B	C	
250													
500													
750													
1000													

Table 3-2E. Alternator Output Versus DC Excitation

Excitation Voltage	Output Phase Voltage*	Output Phase Current (V_p/R)	Output Power per Phase	Total Power
0				
4				
8				
12				
16				
20				

* Take measurements on only one phase.

each current value and the phase voltage for each r/min setting.

D. Complete the current triangle (Fig. 3-2C) and power triangle (Fig. 3-2D) (using power per phase) for the 500-r/min values of Table 3-2F.

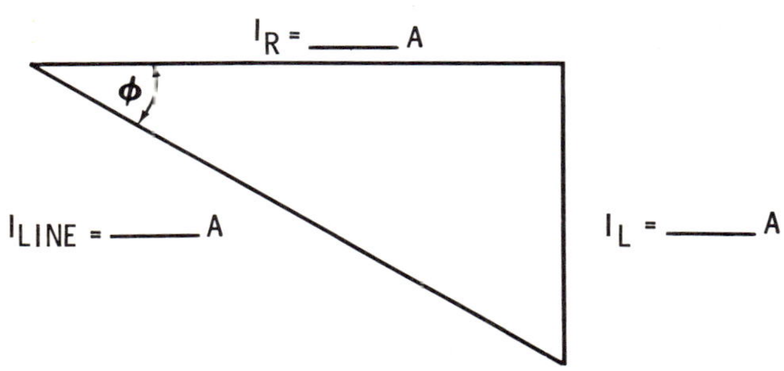

Fig. 3-2C. Current triangle.

7. In this part of the experiment, alternator output will be measured with capacitive loads connected

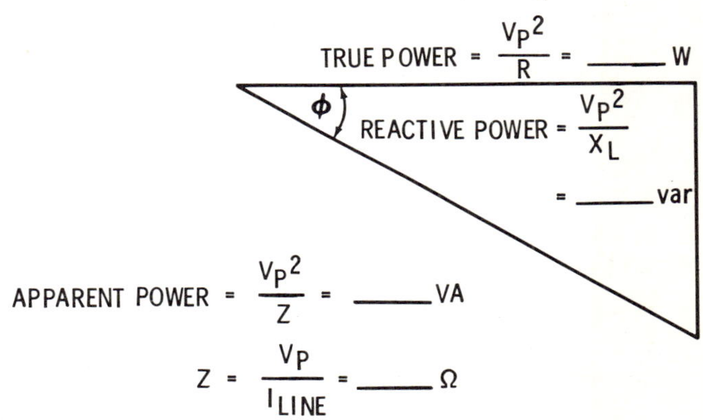

Fig. 3-2D. Power triangle.

Table 3-2F. Alternator Output Values (RL Load)

Speed (r/min)	Line Current (I_{LINE})	Resistive Current (I_R)	Inductive Current (I_L)	Phase Voltage (V_p)	Power per Phase (P_p)	Total Power
250						
500						
750						
1000						

in parallel with the resistive and inductive loads of the previous part of the experiment. The addition of a capacitive load demonstrates "power factor correction."

A. Connect 1.0-μF capacitors in parallel with the lamps and inductors used in the previous part of the experiment. Use only one phase for measurements.

B. Keep the dc excitation at the same value.

C. Complete Table 3-2G using the measured current and voltage values.

D. Complete the current triangle (Fig. 3-2E) and power triangle (Fig. 3-2F) (using power per phase) for the 500-r/min values of Table 3-2G.

Table 3-2G. Alternator Output Values (RLC Load)

Speed (r/min)	Line Current (I_{LINE})	Resistive Current (I_R)	Inductive Current (I_L)	Capacitive Current (I_C)	Phase Voltage (V_p)	Power per Phase (P_p)	Total Power
250							
500							
750							
1000							

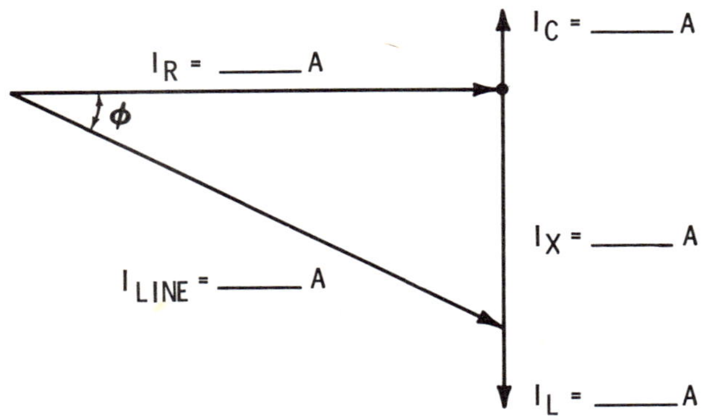

Fig. 3-2E. Current triangle.

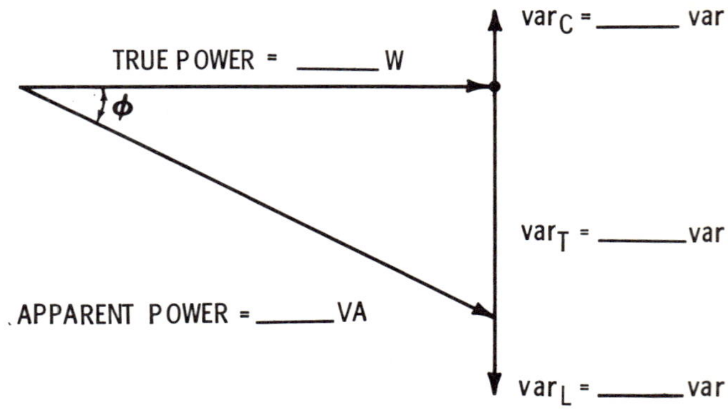

Fig. 3-2F. Power triangle.

8. This concludes the experiment.

Analysis

1. How does an increase in load affect the output values of the three-phase alternator in a delta connection? Wye connection? (See Tables 3-2A, B, C, and D.)

____

____

2. How do the output values of the delta and wye connections of Tables 3-2A and C, and Tables 3-2B and D differ?

____

____

3. From the data of Table 3-2E, how does the dc excitation voltage affect the alternator output?

____

____

4. Compare the current values of Table 3-2F with those of Table 3-2G.

____

____

5. How does the data of Table 3-2G illustrate power factor correction?

____

____

3-3

Characteristics of Chemical Cells

Objective

In this laboratory activity, you will observe some of the characteristics of chemical cells. These dc power sources may be either the primary or secondary type. Those that we will study in this activity are classified as primary cells. Cells may be connected in series to increase voltage output or in parallel to increase current delivery capability. Many times cells are connected in series-parallel combinations to increase both voltage and current levels.

An important characteristic of primary cells is their internal resistance (R_i). As a cell is used over a period of time, its internal resistance becomes greater. Thus, the cell will be limited in the amount of power it can supply to a load. The internal resistance of a cell is usually represented as a resistance in series with the cell. The calculation of a cell's internal resistance will be done in this activity.

Equipment

Multifunction meter.
1.5-V dry cells (3).
Resistors—100 Ω, 1000 Ω.

Procedure

1. Obtain three 1.5-volt dry cells.
2. With the meter, measure the voltage across each dry cell.

   V_1 = ______ volts.

   V_2 = ______ volts.

   V_3 = ______ volts.
3. Connect the cells as shown in each of the five configurations that follow, and measure the voltage from point A to point B.

(1) V_{AB} = ______ volts.

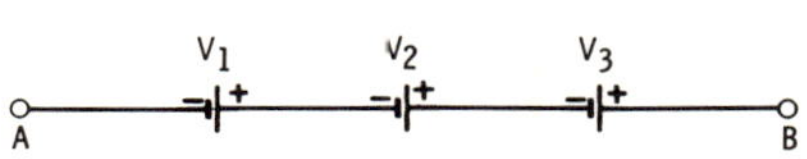

(2) V_{AB} = ______ volts.

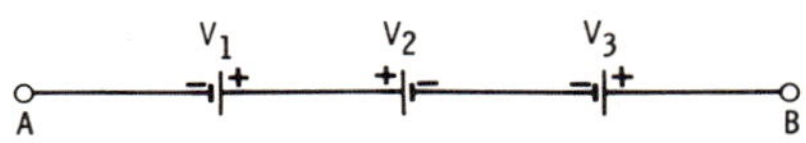

(3) V_{AB} = ______ volts.

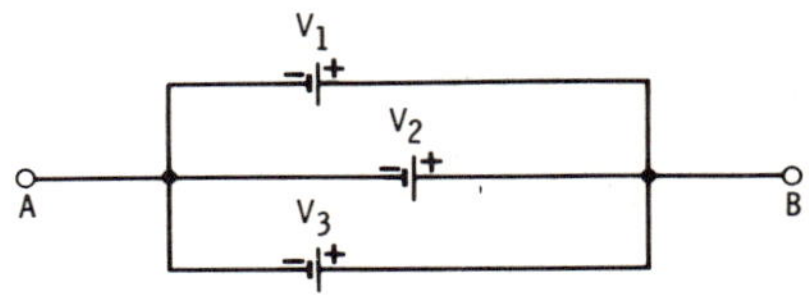

(4) V_{AB} = ______ volts.

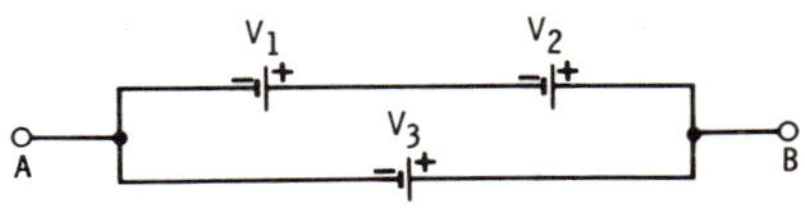

(5) V_{AB} = ______ volts.

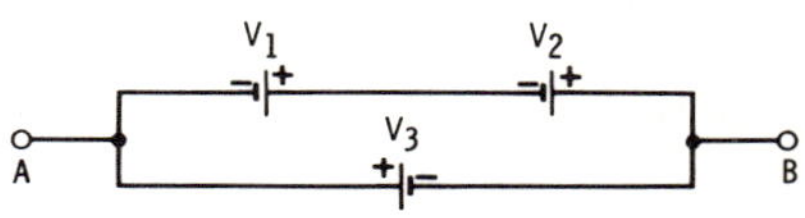

4. To assure that a dry cell is good, one should be checked under a load condition. Measure the voltage across each dry cell while a 100-ohm resistor load is placed across the cell. Record these voltages.

   V_1 = ______ volts.

   V_2 = ______ volts.

   V_3 = ______ volts.
5. Determine the internal resistance (R_i) of each of the dry cells by using the following procedure:

   A. Apply the following formula:

$$R_i = \left(\frac{V_{NL}}{V_{FL}} - 1\right) \times R_L,$$

where V_{NL} is the open-terminal voltage (see Step 2), V_{FL} is the voltage with a load across the cell (see Step 4), and R_L is the value of load resistance.

B. Calculate the internal resistance of each dry cell, using the measured no-load and full-load voltage values.

R_{i_1} = ______ ohms.

R_{i_2} = ______ ohms.

R_{i_3} = ______ ohms.

C. Replace the 100-ohm load with a 1000-ohm load.

D. Measure the voltage across each cell again.

V_1 = ______ volts.

V_2 = ______ volts.

V_3 = ______ volts.

E. Calculate the internal resistance of each dry cell by using the 1000-ohm load.

R_{i_1} = ______ ohms.

R_{i_2} = ______ ohms.

R_{i_3} = ______ ohms.

6. This concludes the experiment.

Analysis

1. Why are chemical cells connected in series?

______

______

2. Why are chemical cells connected in parallel?

______

______

3. Why are chemical cells connected in series-parallel?

______

______

4. Why should a chemical cell always be checked with a load connected?

______

______

5. Compare the voltage readings of Step 4. Did a difference occur? Why?

______

______

6. Discuss each of the voltage measurements obtained in Step 3.

______

______

______

______

______

7. If the open-terminal voltage of a battery is 12.0 volts and its voltage drops to 11.2 volts when a 50-ohm load is added, what is its internal resistance? (Show calculations.)

______

8. From Steps 5B and E, how do the values of internal resistance compare? Why?

______

______

3-4

Permanent-Magnet, Direct-Current Generator

Objective

One method of producing direct-current energy with a rotating machine is the permanent-magnet (pm), direct-current generator. The permanent magnets that are attached to the stator assembly provide the necessary magnetic field. The armature is a wound assembly with a split-ring commutator and brushes from which direct-current energy is delivered to a load.

In this laboratory activity, you will analyze the characteristics of a permanent-magnet dc generator. A pm direct-current motor is constructed the same as the pm dc generator; therefore, a motor could be converted to a generator simply by connecting a prime mover and extracting the voltage from the armature terminals.

Equipment

Prime mover.
Permanent-magnet dc generator (motor).
Multifunction meter.
Tachometer.
Lamps—40 W and 100 W (2).
Oscilloscope.

Procedure

1. A connection diagram for a permanent-magnet dc generator is shown in Fig. 3-4A.
2. Measure the resistance of the armature windings.

   R_A = ______ ohms.
3. Attach a variable-speed prime mover to the shaft of the permanent-magnet dc generator. Start at zero r/min and increase the speed of the prime mover in the increments shown in Table 3-4A. Measure and record the values of dc output voltage and current. Also, calculate the power output for each r/min.
4. Replace the 40-watt lamp load with a 100-watt lamp load.
5. Measure and record the values of dc output voltage and current with the 100-watt load for each r/min increment in Table 3-4B. Calculate the power for each setting.

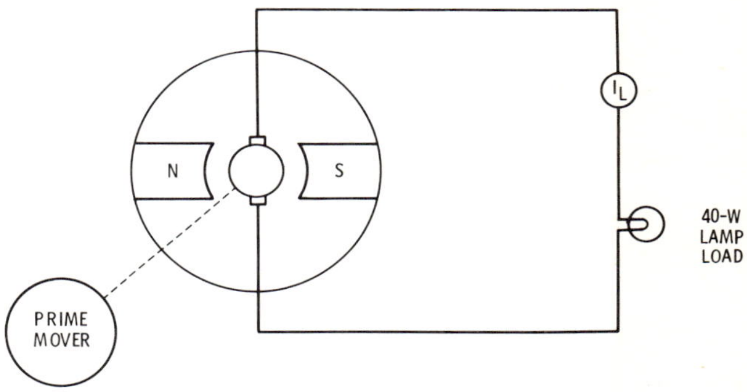

Fig. 3-4A. Connection diagram for a permanent-magnet dc generator.

Table 3-4A. Permanent-Magnet DC Generator Values (40-W Load)

r/min	Output Voltage (V_o)	Load Current (I_L)	Power ($V_o \times I_L$)
500			
750			
1000			
1250			
1500			

6. With the generator rotating at 1000 r/min, momentarily connect another 100-watt lamp across the output. Note the effect on speed.

   Speed = ______ r/min.

Table 3-4B. Permanent-Magnet DC Generator Values (100-W Load)

r/min	Output Voltage	Load Current	Power Output
500			
750			
1000			
1250			
1500			

7. Connect an oscilloscope across the load and record the waveform developed at 1000 r/min. Record this waveform on the following graph form:

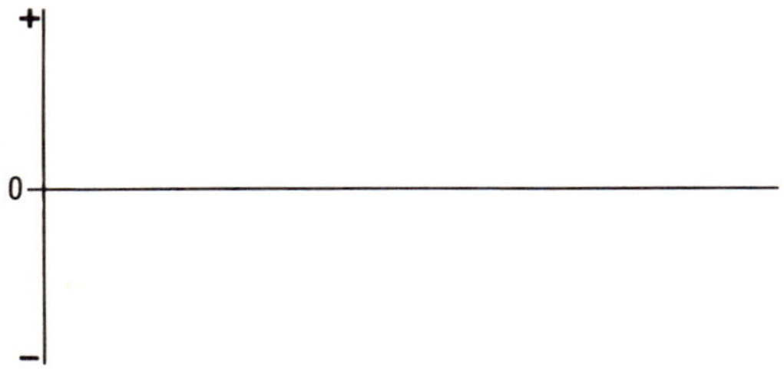

Analysis

1. What are some advantages of permanent-magnet dc generators?

____

____

2. Compare each of the following values from Tables 3-4A and B at the 750-r/min setting:
   (a) Output voltage.

   ____

   (b) Load current.

   ____

   (c) Power output.

   ____

3. Make a graph showing the relationship of voltage output (vertical axis) and load current (horizontal axis) for each load value. Put both curves on the following graph form:

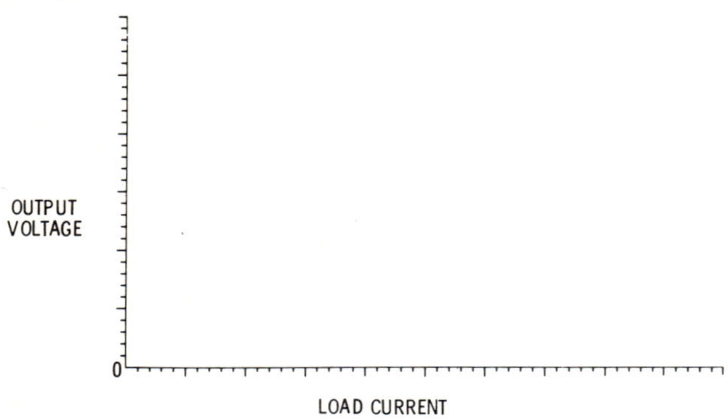

4. How do the following factors vary in a permanent-magnet dc generator?
   (a) As r/min increases, voltage output

   ____________.

   (b) As r/min increases, load current

   ____________.

   (c) As r/min increases, the power output

   ____________.

5. How is the speed of the permanent-magnet dc generator affected by load (see Step 6)?

____

6. What type of waveform was observed in Step 7?

____

7. Discuss the basic operation of a permanent-magnet dc generator.

____

____

____

____

3-5

Separately Excited, Direct-Current Generator

Objective

The separately excited dc generator obtains its field excitation from a direct-current source. Usually this dc source is variable, so that the output of the generator can be adjusted. Very precise voltage outputs over a wide range of loads may be obtained with this type of generator.

In this laboratory activity, you will analyze the operational characteristics of a separately excited dc generator. If this type of machine is not available in the lab, it is possible to modify many types of dc motors to operate as a separately excited generator. A variable dc voltage is connected to its shaft, and the output is taken directly across the armature terminals at the brushes. A unit constructed in the lab using a rotating machinery unit is desirable to analyze.

Equipment

Prime mover.
Separately excited dc generator.
Multifunction meter.
Tachometer.
Lamps—40 W and 100 W (2).
Variable dc power source.
Oscilloscope.

Procedure

1. A connection diagram for a separately excited dc generator is shown in Fig. 3-5A.

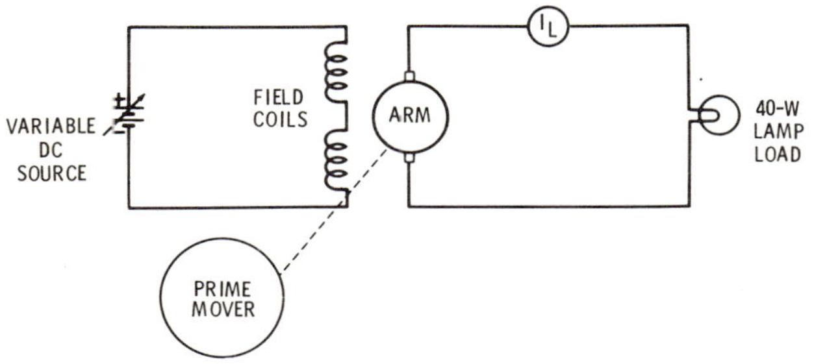

Fig. 3-5A. Connection diagram for a separately excited dc generator.

2. Measure the resistance of the armature and field windings.

Table 3-5A. Separately Excited DC Generator Values (40-W Load)

r/min	Output Voltage (V_o)	Load Current (I_L)	Power ($V_o \times I_L$)
500			
750			
1000			
1250			
1500			

R_A = ______ ohms.

R_F = ______ ohms.

3. Apply the rated dc voltage to the field windings. Attach a variable-speed prime mover to the shaft of the separately excited dc generator. Start at zero r/min and increase the speed of the prime mover in the increments shown in Table 3-5A. Measure and record the values of dc output voltage and current. Also, calculate the power output for each r/min.
4. Replace the 40-watt lamp load with a 100-watt lamp load.
5. Measure and record the values of dc output voltage and current with the 100-watt load for each r/min increment in Table 3-5B. Calculate the power for each setting.
6. With the generator rotating at 1000 r/min, momentarily connect another 100-watt lamp across the output. Note the effect on speed.

Speed = ______ r/min.

Table 3-5B. Separately Excited DC Generator Values (100-W Load)

r/min	Output Voltage	Load Current	Power Output
500			
750			
1000			
1250			
1500			

7. Connect an oscilloscope across the load and record the waveform developed at 1000 r/min. Record this waveform on the following graph form:

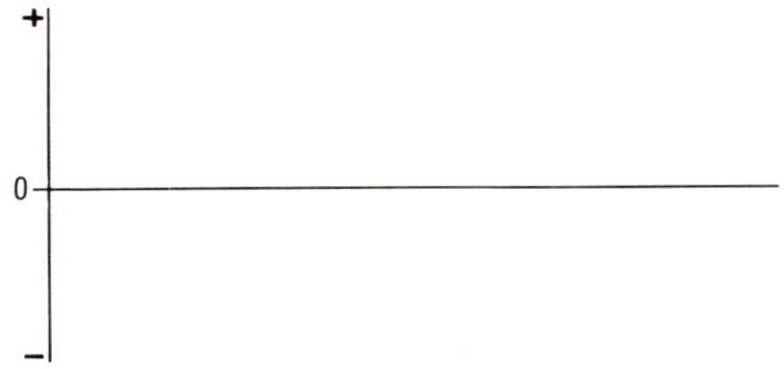

8. Connect a 40-watt lamp load across the output of the generator. Adjust the dc excitation voltage to zero volts. Complete Table 3-5C by adjusting the dc excitation voltage in the increments shown. Keep the speed constant at 1000 r/min.

Table 3-5C. DC Excitation Values With 40-W Load

Excitation Voltage	Output Voltage	Load Current	Power Output
0			
2			
4			
6			
8			
10			

Analysis

1. What are some characteristics of separately excited dc generators?

____

____

2. From the data of Table 3-5C, what effect does an increase in dc excitation have on:
   (1) Output voltage?
   (2) Load current?
   (3) Power output?

____

____

3. Make a graph showing the relationship of voltage output (vertical axis) and load current (horizontal axis) for each load value. Put both curves on the following form:

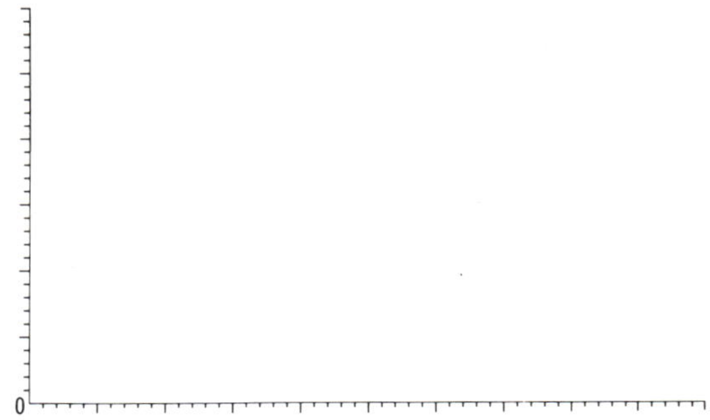

4. How do the following factors vary in a separately excited dc generator?
   (a) As r/min increases, voltage output
   ______________.
   (b) As r/min increases, load current
   ______________.
   (c) As r/min increases, power output
   ______________.

5. How is the speed of the separately excited dc generator affected by load (see Step 6)?

____

____

6. Compare the armature and field resistances from Step 2.

____

____

3-6

Series-Wound, Direct-Current Generator

Objective

The series-wound dc generator is constructed in the same way as a series-wound dc motor (or universal ac/dc motor). The most undesirable characteristic of this type of generator is its unstable voltage output as load changes. Since, as load current increases, the field current increases by the same amount, a higher output will result. Therefore, voltage output level is entirely dependent upon the magnitude of the load current.

In this laboratory activity, you will analyze the characteristics of a commercially manufactured series-wound dc generator or one which has been constructed in the lab. A universal motor could be operated as a generator to complete this analysis.

Equipment

Variable-speed prime mover.
Series-wound dc generator.
Multifunction meter.
Tachometer.
Lamps—40 W and 100 W (2).
Oscilloscope.

Procedure

1. A connection diagram for a series-wound dc generator is shown in Fig. 3-6A.

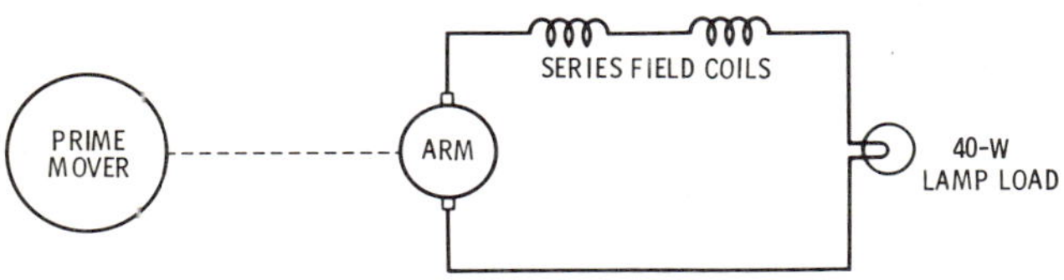

Fig. 3-6A. Connection diagram for a series-wound dc generator.

2. Measure the resistance of the armature and field windings.

   R_A = ______ ohms.

   R_F = ______ ohms.
3. Attach a variable-speed prime mover to the shaft of the series-wound dc generator. Start at zero r/min and increase the speed of the prime mover in the increments shown in Table 3-6A. It may be necessary to "flash the field" of the generator. Measure and record the values of dc output voltage and current. Also, calculate the power output for each r/min.

Table 3-6A. DC Series-Wound Generator Values (40-W Load)

r/min	Output Voltage (V_o)	Load Current (I_L)	Power ($V_o \times I_L$)
500			
750			
1000			
1250			
1500			

4. Replace the 40-watt lamp load with a 100-watt lamp load.
5. Measure and record the values of dc output voltage and current with the 100-watt load for each r/min increment in Table 3-6B. Calculate the power for each setting.
6. With the generator rotating at 1000 r/min, momentarily connect another 100-watt lamp across the output. Note the effect on speed.

   Speed = ______ r/min.

Table 3-6B. DC Series-Wound Generator Values (100-W Load)

r/min	Output Voltage	Load Current	Power Output
500			
750			
1000			
1250			
1500			

7. Connect an oscilloscope across the load and record

the waveform developed at 1000 r/min. Record this waveform on the following graph form:

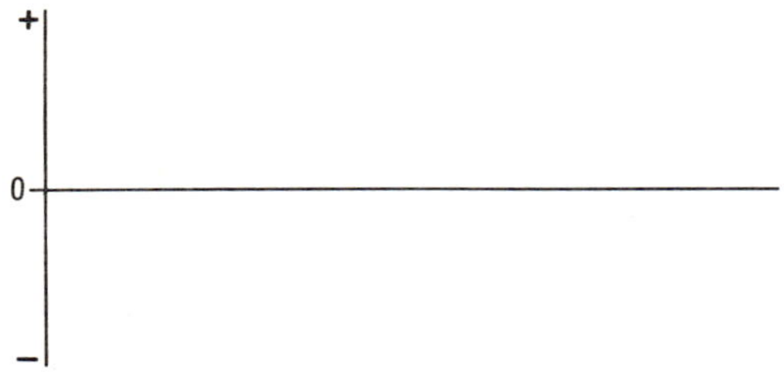

Analysis

1. Compare each of the following values from Tables 3-6A and B at the 1000 r/min setting:
   (a) Output voltage.

   (b) Load current.

   (c) Power output.

2. Make a graph showing the relationship of voltage output (vertical axis) and load current (horizontal axis) for each load value. Put both curves on the following graph form:

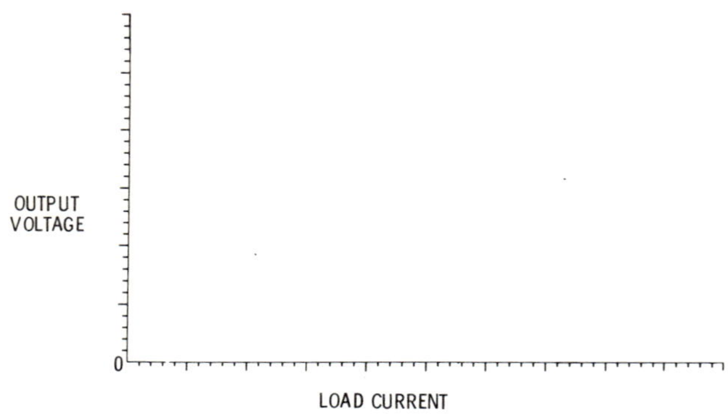

3. How do the following factors vary in a dc series generator?
   (a) As r/min increases, voltage output ________.
   (b) As r/min increases, load current ________.
   (c) As r/min increases, power output ________.

4. How is the speed of the series-wound dc generator affected by load (see Step 6)?

5. What type of waveform was observed in Step 7?

6. Compare the armature and field resistances from Step 2 of the series dc generator.

7. Compare the armature and field resistances of series and shunt dc generators.

8. What is meant by "flashing the field" of a self-excited dc generator?

3-7

Shunt-Wound, Direct-Current Generator

Objective

The shunt-wound dc generator is usually more desirable as a dc power source than its series-wound counterpart. The field windings of this machine have a comparatively high resistance; therefore, only a small current flows in the field regardless of the load value. The voltage output of the shunt machine is quite stable compared to that of the series-wound dc generator.

In this laboratory activity, you will analyze the operational characteristics of a shunt-wound dc generator. Either a commercially manufactured shunt-wound generator or a laboratory unit may be used.

Equipment

Variable-speed prime mover.
Shunt-wound dc generator.
Multifunction meter.
Tachometer.
Lamps—40 W and 100 W (2).
Oscilloscope.

Procedure

1. A connection diagram for a shunt-wound dc generator is shown in Fig. 3-7A.

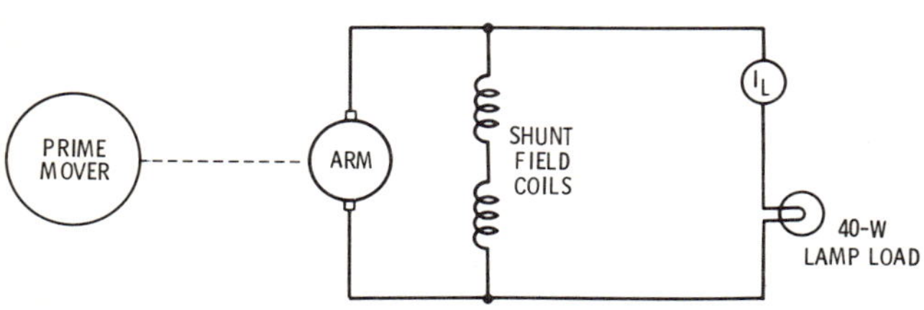

Fig. 3-7A. Connection diagram for a shunt-wound dc generator.

2. Measure the resistance of the armature and field windings.

   R_A = ______ ohms.

   R_F = ______ ohms.

3. Attach a variable-speed prime mover to the shaft of the shunt-wound dc generator. Start at zero r/min and increase the speed of the prime mover in the increments shown in Table 3-7A. It may be necessary to "flash the field" of the generator. Measure and record the values of dc output voltage and current. Also, calculate the power output for each r/min.

Table 3-7A. DC Shunt-Wound Generator Values (40-W Load)

r/min	Output Voltage (V_o)	Load Current (I_L)	Power ($V_o \times I_L$)
500			
750			
1000			
1250			
1500			

4. Replace the 40-watt lamp load with a 100-watt lamp load.
5. Measure and record the values of dc output voltage and current with the 100-watt load for each r/min increment in Table 3-7B. Calculate the power for each setting.

Table 3-7B. DC Shunt-Wound Generator Values (100-W Load)

r/min	Output Voltage	Load Current	Power Output
500			
750			
1000			
1250			
1500			

6. With the generator rotating at 1000 r/min, momentarily connect another 100-watt lamp across the output. Note the effect on speed.
7. Connect an oscilloscope across the load and record

the waveform developed at 1000 r/min. Record this waveform on the following graph form:

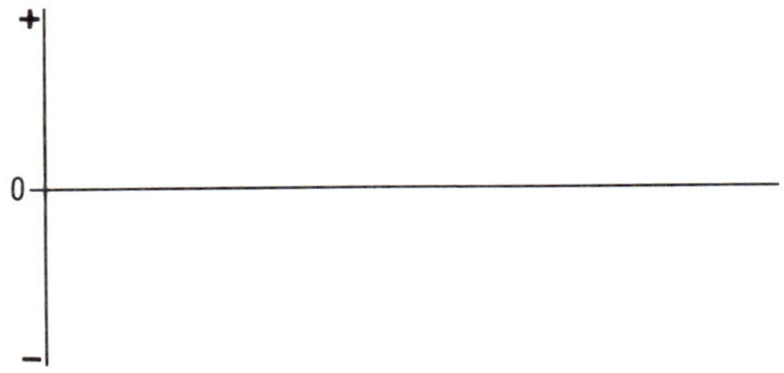

Analysis

1. What are some applications of dc generators?

2. Compare each of the following values from Tables 3-7A and B at the 1000 r/min setting:
   (a) Output voltage.
   (b) Load current.
   (c) Power output.

3. Make a graph showing the relationship of voltage output (vertical axis) and load current (horizontal axis) for each load value. Put both curves on the following graph form:

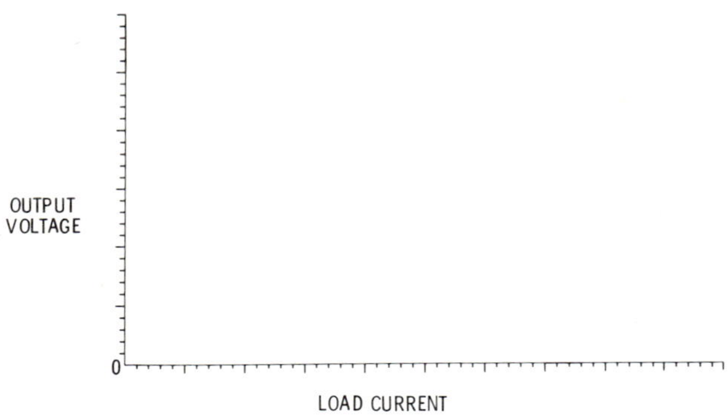

4. How do the following factors vary in a dc shunt generator?
   (a) As r/min increases, voltage output ________.
   (b) As r/min increases, load current ________.
   (c) As r/min increases, power output ________.

5. How is the speed of a dc generator affected by load (see Step 6)?

6. What type of waveform was observed in Step 7?

7. How does the compound-wound dc generator differ from the shunt generator?

8. Compare the armature and field resistances from Step 2.

Semiconductor Diode Characteristics

Objective

The semiconductor diode is one of the most widely used of all electrical components. One of the primary functions of diode devices is rectification. Diode rectification will be studied in some of the laboratory activities that follow. A knowledge of diode characteristics is needed before rectifier circuits can be fully understood.

In this laboratory activity, you will observe the forward and reverse bias characteristics of a pn junction semiconductor diode. A look at these characteristics will allow you to see how a diode can act as a rectifier for conversion of ac to dc.

Equipment

Semiconductor diode—1N4004 or equivalent.
Electronic multifunction meter.
Variable dc power source.
Resistor—1000 Ω.

Procedure

1. Construct the diode characteristics experimental circuit shown in Fig. 3-8A. This is a forward-bias connection.

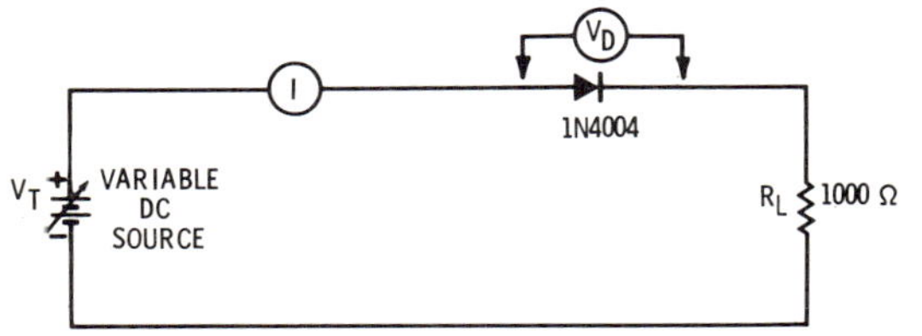

Fig. 3-8A. Circuit for observing semiconductor diode forward-bias characteristics.

2. Before connecting the diode into the circuit, check its forward and reverse resistances with a meter. This gives a quick indication of the diode's condition.

   Forward resistance = ______ ohms.

   Reverse resistance = ______ ohms.

3. Connect the diode into the circuit. Apply power and measure V_{R_L}, V_D, and I. Measure these values as the source voltage (V_T) is varied in the increments shown in Table 3-8A. Complete Table 3-8A.

Table 3-8A. Forward Bias Characteristics

V_T	V_{R_L}	V_D	I
0			
1			
2			
5			
10			
15			
20			

4. Modify the circuit as shown in Fig. 3-8B. This is the reverse-bias connection.

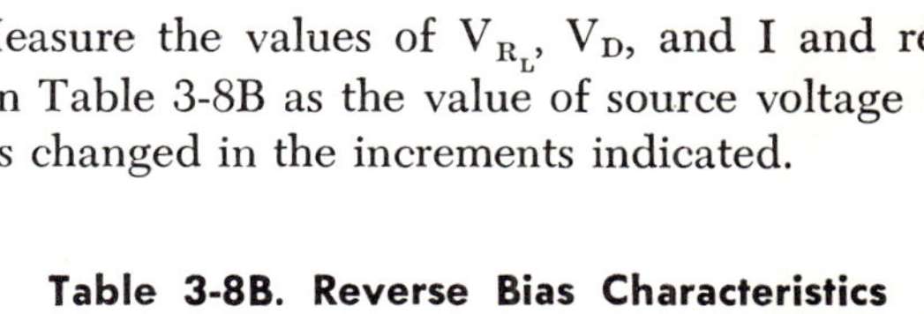

Fig. 3-8B. Circuit for observing semiconductor diode reverse-bias characteristics.

5. Measure the values of V_{R_L}, V_D, and I and record in Table 3-8B as the value of source voltage (V_T) is changed in the increments indicated.

Table 3-8B. Reverse Bias Characteristics

V_T	V_{R_L}	V_D	I
0			
5			
10			
15			
20			

6. This concludes the experiment.

Analysis

1. Why does the semiconductor diode conduct in the forward-bias connection but not in the reverse-bias connection?

2. What are some advantages of semiconductor diodes over vacuum-tube diodes and gaseous-tube diodes?

3. Define the following terms:
   (a) Potential barrier.

   (b) Reverse voltage breakdown.

   (c) Forward current rating.

   (d) Peak reverse voltage (prv) rating.

4. Using the data from Tables 3-8A and B, construct the following graph, which represents the forward and reverse characteristics of the 1N4004 diode:

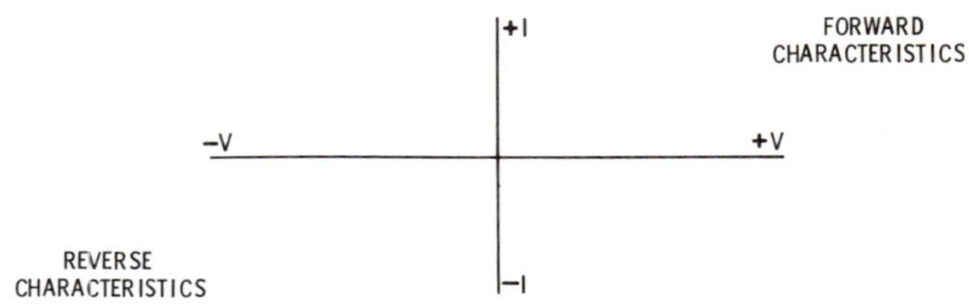

Semiconductor Diode, Single-Phase, Half-Wave Rectifier

Objective

In this laboratory activity, you will observe the characteristics of a single-phase, half-wave rectification system that uses a semiconductor diode. Most industrial applications that require direct-current power sources now use semiconductor rectifying devices. There are still certain applications that use gaseous-tube diodes (called *phanotrons*) or vacuum-tube diodes; however, their uses are now much surpassed by the semiconductor diode.

The rectification method used in this experiment is only one method to convert alternating current into direct current. In the following experiments, you will investigate other methods of rectification. The semiconductor diode, single-phase, half-wave rectifier is the simplest method that could be used for ac to dc conversion.

Equipment

Variable ac power supply.
Electronic multifunction meter.
Resistor—1000 Ω.
Diode—1N4004 or equivalent.
Oscilloscope.

Procedure

1. Construct the single-phase, half-wave rectifier circuit shown in Fig. 3-9A.
2. With a voltmeter, measure the ac values across D_1 and R_L, and record in Table 3-9A, using 10 volts ac as the applied voltage.
3. Adjust the ac power source to 20 volts ac and measure V_{D_1} and V_{R_L}. Record your values in Table 3-9A.

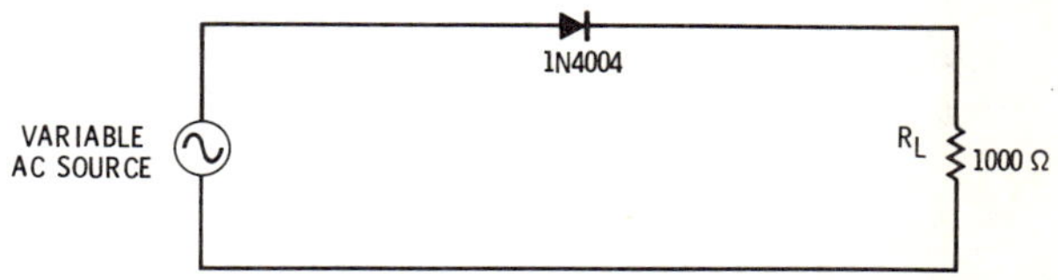

Fig. 3-9A. Semiconductor diode, single-phase, half-wave rectifier circuit.

4. Adjust the ac power source to 30 volts ac and measure V_{D_1} and V_{R_L}. Record your values in Table 3-9A.
5. Peak voltage is equal to 1.41 × rms voltage. Calculate the peak values of V_{in}, V_{D_1}, and V_{R_L} and record in Table 3-9A.
6. Prepare the meter to measure dc voltage and complete the dc voltage values in Table 3-9B for 10, 20, and 30 volts ac applied. (V_{in} is the dc voltage measurement across the ac source.)

Table 3-9B. DC Values

Applied Voltage	V_{in} (dc)	V_{D_1}	V_{R_L}
10 V ac			
20 V ac			
30 V ac			

7. Apply 10 volts ac to the circuit. With an oscilloscope properly synchronized with the ac input, observe and record the necessary waveforms in Table 3-9C.
8. Also, record the V_{peak} and $V_{p\text{-}p}$ values in Table 3-9C.
9. This concludes the experiment.

Table 3-9A. AC Values

Applied Voltage	V_{in} (peak)	V_{D_1} (rms)	V_{D_1} (peak)	V_{R_L} (rms)	V_{R_L} (peak)
10 V ac					
20 V ac					
30 V ac					

Table 3-9C. Oscilloscope Waveforms

		V_{peak}	$V_{p\text{-}p}$
V_{in}	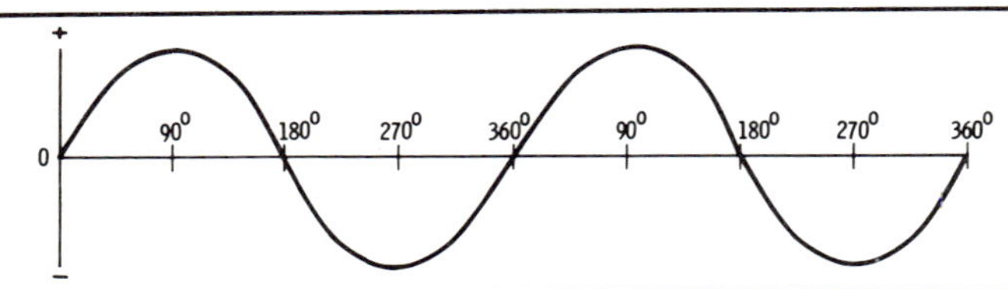		
V_{D_1}	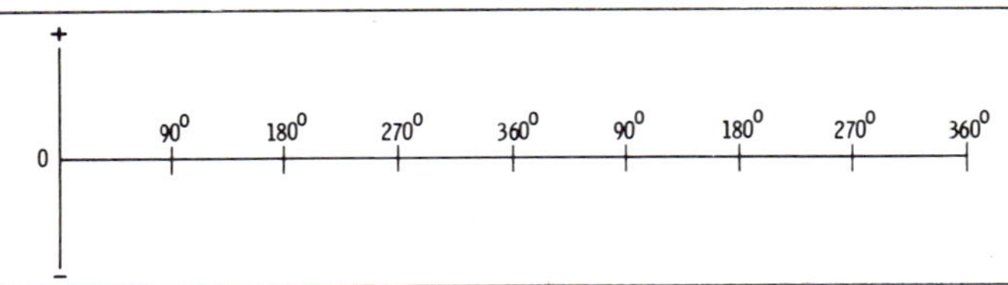		
V_{R_L}	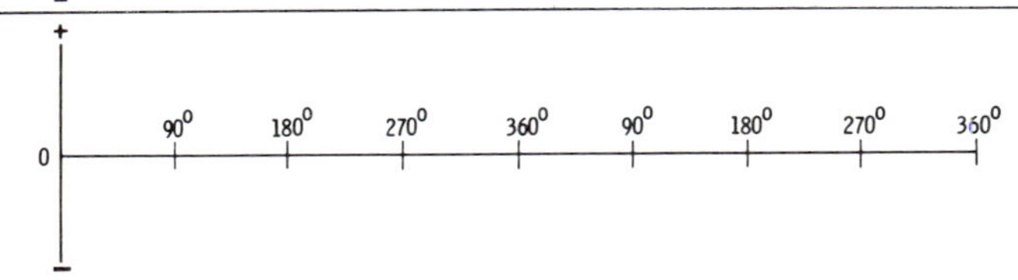		

Analysis

1. What is the frequency of the ac ripple voltage across R_L?

____

2. What effect would a lower value of R_L have on the operation of this rectifier circuit?

____

3. What is the ratio of dc voltage output to peak ac voltage input for this rectifier circuit?

$$\frac{V_{dc(out)}}{V_{peak(in)}} = \_\_\_\_\_\_.$$

4. What is the formula used for determining the value of dc voltage output for a semiconductor diode, single-phase, half-wave rectifier?

5. Draw a diagram showing how a vacuum-tube diode could be used in place of D_1 in this rectifier circuit.

6. What effect does increasing the applied ac voltage have on the dc output of this rectifier circuit?

____

7. What would happen if the polarity of D_1 was reversed?

____

3-10

Vacuum-Tube Diode, Single-Phase, Half-Wave Rectifier

Objective

In this laboratory activity, you will observe the operation of a single-phase, vacuum-tube rectifier circuit. This type of rectifier is almost obsolete; however, this activity demonstrates the operational principles of a vacuum-tube diode.

In the vacuum-tube diode, 6.3 volts ac is applied to the filament pins of the tube. The current through the filament produces heat, which affects the cathode that is placed near the filament inside the tube enclosure. The cathode material emits negatively charged electrons when it is heated. These electrons form a space charge around the cathode. When a positive potential is applied to the plate, the electrons will flow toward the plate. This constitutes a current. Therefore, by varying the positive potential on the plate of the tube, current flow to an external load may be varied. When a negative potential is applied to the plate, electrons are repelled back to the cathode. Therefore, no current will flow. This principle is used in vacuum-tube rectifier circuits for ac to dc conversion.

Equipment

Electronic multifunction meter.
Vacuum tube—6X4 or equivalent.
6.3 V ac source.
Resistor—10,000 Ω.
Variable ac source.

Procedure

1. Construct the vacuum-tube diode, single-phase, half-wave rectifier circuit shown in Fig. 3-10A.

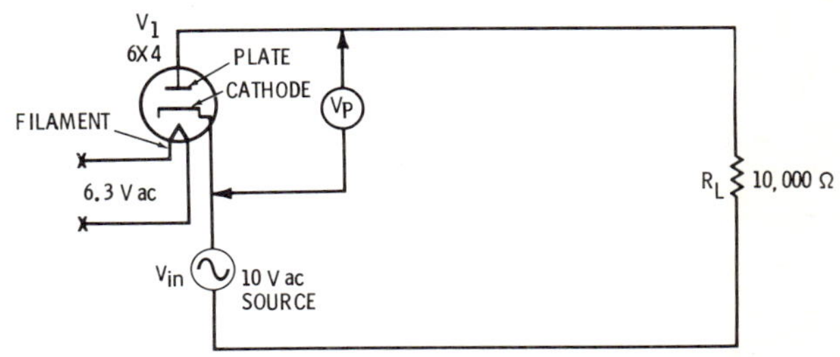

Fig. 3-10A. Vacuum-tube diode, single-phase, half-wave rectifier circuit.

2. With a voltmeter, measure the ac values across V_1 and R_L, and record in Table 3-10A, using 10 volts ac as the ac input voltage.
3. Adjust the ac power source to 20 volts ac and measure V_p and V_{R_L}. Record your values in Table 3-10A.
4. Adjust the ac power source to 30 volts ac and measure V_p and V_{R_L}. Record your values in Table 3-10A.
5. Peak voltage is equal to 1.41 × rms voltage. Calculate the peak values of V_{in}, V_p, and V_{R_L} and record in Table 3-10A.
6. Prepare the meter to measure dc voltage and complete the dc voltage values in Table 3-10B for 10, 20, and 30 volts ac applied. (V_{in} is the dc voltage measurement across the ac source.)

Table 3-10B. DC Values

Applied Voltage	V_{in} (dc)	V_p	V_{R_L}
10 V ac			
20 V ac			
30 V ac			

Table 3-10A. AC Values

Applied Voltage	V_{in} (peak)	V_p (rms)	V_p (peak)	V_{R_L} (rms)	V_{R_L} (peak)
10 V ac					
20 V ac					
30 V ac					

Table 3-10C. Oscilloscope Waveforms

		V_{peak}	$V_{p\text{-}p}$
V_{in}	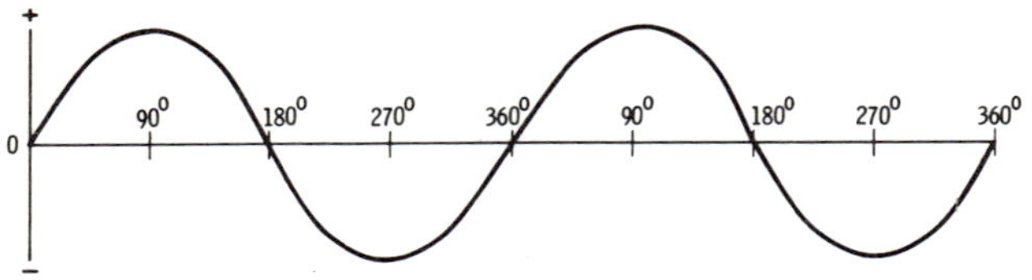		
V_p	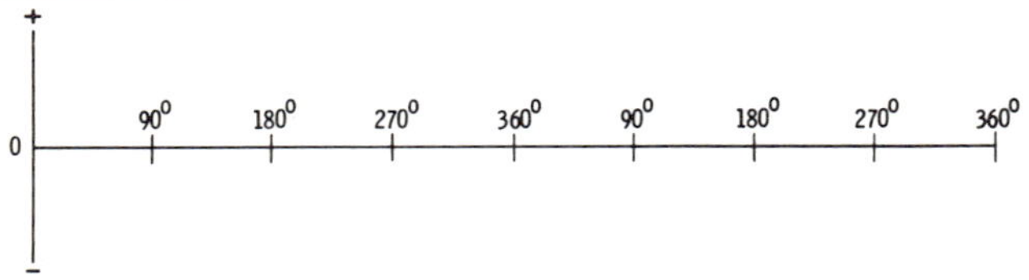		
V_{R_L}	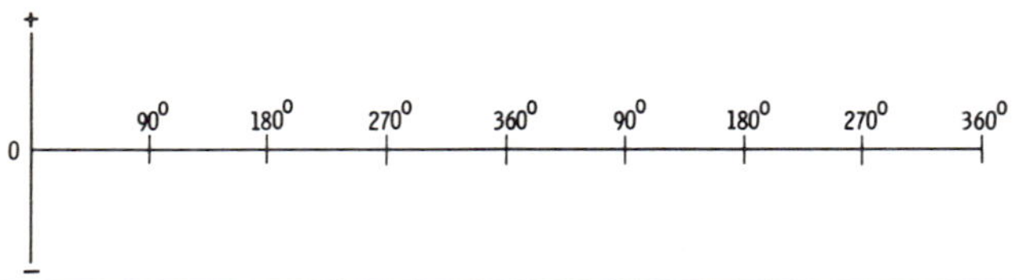		

7. Apply 10 volts ac to the circuit. With an oscilloscope properly synchronized with the ac input, observe and record the necessary waveforms in Table 3-10C.
8. Also, record the V_{peak} and $V_{p\text{-}p}$ values in Table 3-10C.
9. This concludes the experiment.

Analysis

1. What is the frequency of the ac ripple voltage across R_L?

____

2. What is the ratio of dc voltage output to peak ac voltage input for the vacuum-tube rectifier circuit you constructed?

$$\frac{V_{dc(out)}}{V_{peak(in)}} = \underline{\qquad\qquad}.$$

3. What effect does increasing the applied ac voltage have on the dc output of this rectifier?

____

4. How does the ratio of $V_{dc(out)}$ to $V_{peak(in)}$ for this rectifier compare to that of the semiconductor half-wave rectifier?

____

5. What are some disadvantages of this rectifier circuit compared to the semiconductor rectifier?

____

____

____

3-11

Semiconductor Diode, Single-Phase, Full-Wave Rectifier

Objective

In this laboratory activity, you will observe the characteristics of a single-phase, full-wave rectification system that uses semiconductor diodes and a center-tapped transformer. A standard 6.3/12.6-volt ac filament transformer will be used.

Full-wave rectifiers have an advantage over their half-wave counterparts of a higher dc voltage output. However, the method observed in this experiment requires two diode devices and a relatively expensive center-tapped transformer.

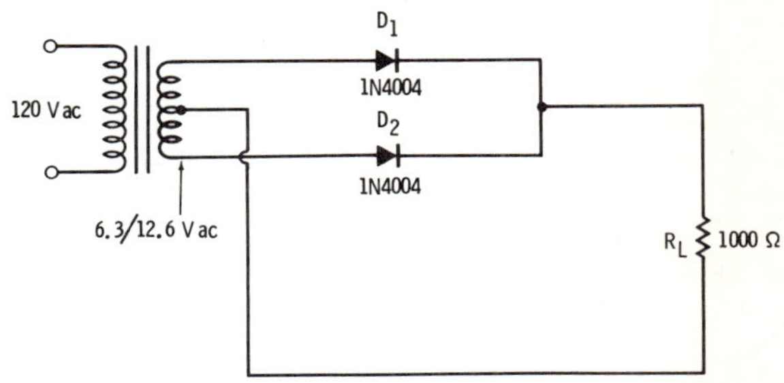

Fig. 3-11A. Semiconductor diode, single-phase, full-wave rectifier circuit.

Equipment

Center-tapped 6.3/12.6-V ac transformer.
Electronic multifunction meter.
Resistor—1000 Ω.
Diodes (2)—1N4004 or equivalent.
Oscilloscope.

Procedure

1. Construct the circuit shown in Fig. 3-11A.
2. With a voltmeter, measure the ac voltage values across D_1, D_2, and R_L and record in Table 3-11A.
3. Calculate the peak values of V_{In}, V_{D_1}, V_{D_2}, and V_{R_L} and record in Table 3-11A.
4. Prepare the meter to measure dc voltage and complete the dc values in Table 3-11B.
5. With an oscilloscope properly synchronized with the ac input, observe and record the necessary waveforms in Table 3-11C.
6. Also, record V_{peak} and $V_{p\text{-}p}$ values in Table 3-11C.
7. This concludes the experiment.

Analysis

1. What is the frequency of the ac ripple voltage across R_L?

   ____

2. Would using a lower value resistance for R_L have the same effect as on the single-phase, half-wave rectifier? Why?

   ____

   ____

   ____

3. What is the ratio of dc voltage output to peak ac voltage input for the rectifier constructed in this experiment?

   $$\frac{V_{dc(out)}}{V_{peak(in)}} = \_\_\_\_\_\_\_\_\_\_.$$

4. What is the formula for determining the value of dc voltage output for a single-phase, full-wave rectifier?

Table 3-11A. AC Values

V_{In}	V_{In} (peak)	V_{D_1} (peak)	V_{D_2}	V_{D_2} (peak)	V_{R_L}	V_{R_L} (peak)
12.6 V ac						

Table 3-11B. DC Values

V_{In}	V_{In} (dc)	V_{D_1}	V_{D_2}	V_{R_L}
12.6 V ac				

Table 3-11C. Oscilloscope Waveforms

		V_{peak}	$V_{p\text{-}p}$
V_{in}	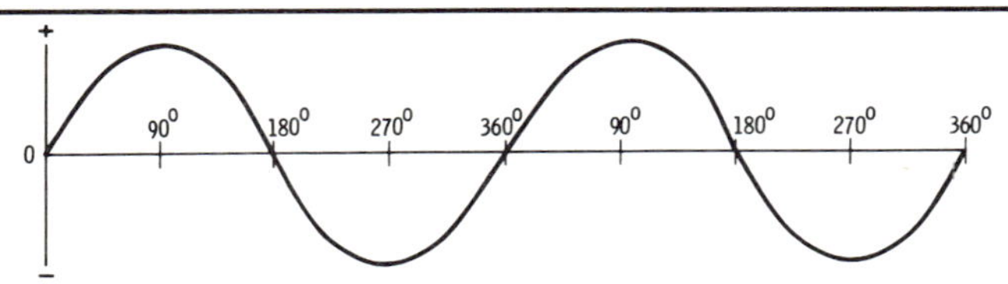		
V_{D_1}	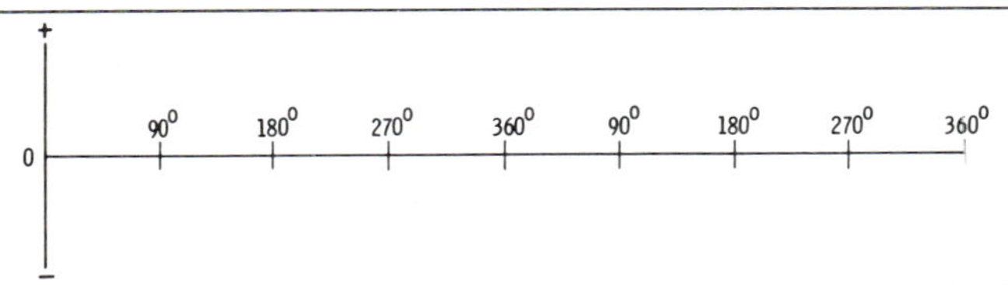		
V_{D_2}	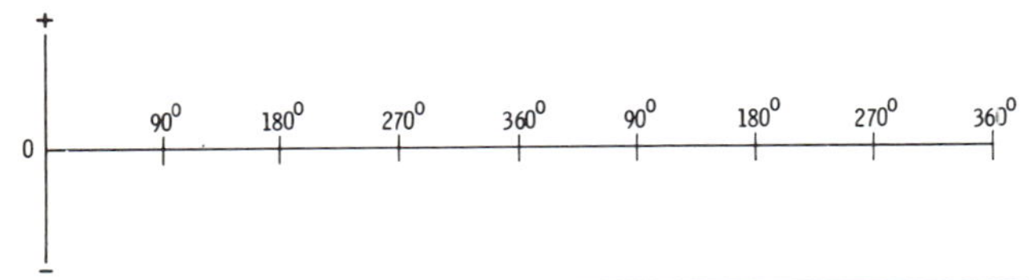		
V_{R_L}	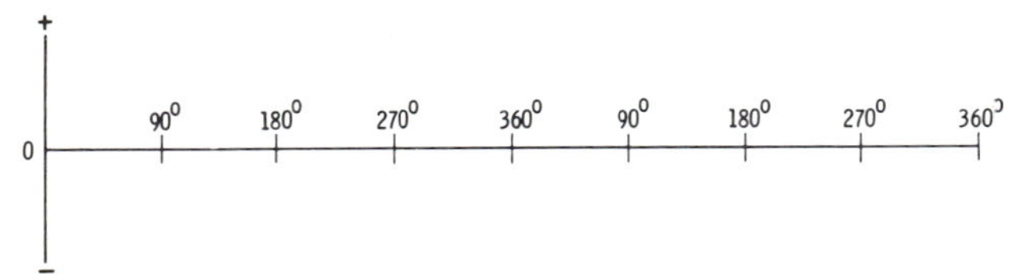		

5. Draw a diagram showing how two vacuum-tube diodes could be used to produce full-wave rectification.

6. What would happen if the polarities of both D_1 and D_2 were reversed?

____

3-12

Gaseous-Tube Diode, Single-Phase, Full-Wave Rectifier

Objective

Gaseous-tube diodes are used as high-current rectifiers and regulators because they are efficient, have a low internal voltage drop, and can handle large amounts of current. In this activity, you will construct a gaseous-tube diode, full-wave rectifier that is used in heavy-duty power supplies.

Equipment

Gaseous-tube diodes (2)—866A.
Transformer—120-V primary; 320-V center-tapped secondary.
Transformer—120-V primary; 2.5-V center-tapped secondary.
Multifunction meter.
Resistors—10,000 Ω, 5000 Ω, 1000 Ω (2), 5 W.
Capacitor—30 μF, 450 V dcw.
Regulator tube—OA2.

Procedure

1. Construct the gaseous-tube diode, full-wave rectifier circuit shown in Fig. 3-12A.
2. With a voltmeter, measure the ac voltage values across tube V_1, tube V_2, and R_L. Record these values in Table 3-12A.
3. Calculate the peak values of V_{in}, V_1, V_2 and V_{R_L} and record in Table 3-12A.
4. Prepare the meter to measure dc voltage and complete the dc values in Table 3-12B.
5. Modify the circuit as shown in Fig. 3-12B.

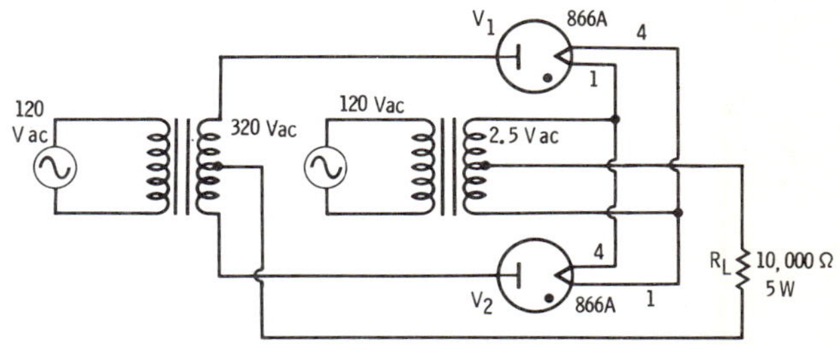

Fig. 3-12A. Gaseous-tube diode, single-phase, full-wave rectifier circuit.

6. Perform the measurements necessary to complete Table 3-12C as the load resistance is changed from 10,000 ohms to 1000 ohms.

Table 3-12C. Load Values

Load	Output Voltage (V_{R_L})	Load Current (I_L)	Total Current (I_T)
10,000 Ω			
5000 Ω			
1000 Ω			

7. This concludes the experiment.

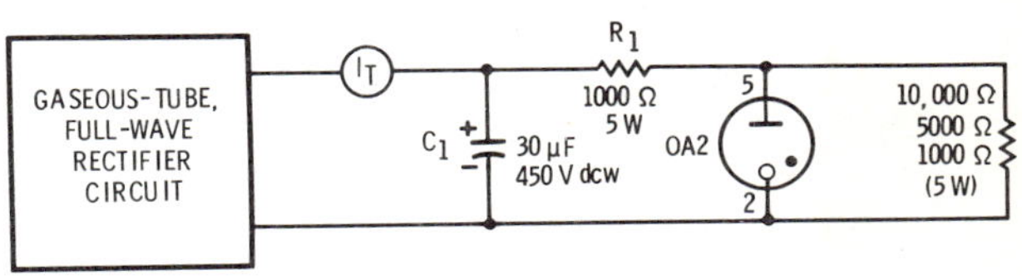

Fig. 3-12B. Modified gaseous-tube, full-wave rectifier circuit.

Table 3-12A. AC Values

V_{in}	V_{in} (peak)	V_1	V_2	V_1 (peak)	V_2 (peak)	V_{R_L}
320 V ac						

Table 3-12B. DC Values

V_{in} (ac)	V_{in} (dc)	V_1	V_2	V_{R_L}
320 V ac				

Analysis

1. What is the ratio of dc voltage output to peak ac voltage input for the rectifier constructed in this experiment?

$$\frac{V_{dc(out)}}{V_{peak(in)}} = \underline{\hspace{3cm}}.$$

2. Briefly explain the operation of the gaseous tube used in this activity.

____

____

____

____

3. What are some applications where gaseous-tube rectifiers might be used in industry?

____

____

____

4. In Step 6, what change in output voltage occurred in comparison with the data of Table 3-12B?

____

____

5. Calculate the voltage regulation of the gaseous-tube rectifier power supply using the data of Table 3-12C:

$$\% \text{ Regulation}^* = \frac{V_{NL} - V_{FL}}{V_{FL}} \times 100 = \underline{\hspace{2cm}}\%.$$

*Use V_{NL} = 10,000-ohm load value and V_{FL} = 1000-ohm load value.

6. What effect does a smaller value of load resistance (increased load) have on load current? Total current?

____

____

Single-Phase, Full-Wave Bridge Rectification

Objective

Full-wave rectification systems are more desirable from an operational point of view than half-wave systems are. It is possible to accomplish full-wave rectification by using a center-tapped transformer winding and two diode devices. However, the transformer is relatively expensive.

Full-wave rectification is also accomplished by using four diodes connected in a "bridge" configuration. This circuit eliminates the need for a center-tapped transformer. In this laboratory activity, you will observe the operation of a single-phase bridge rectification system. Such systems are commonly used for converting alternating current to direct current.

Equipment

Variable ac power supply.
Electronic multifunction meter.
Resistor—10,000 Ω.
Diodes (4)—1N4004 or equivalent.
Oscilloscope.

Procedure

1. Construct the circuit shown in Fig. 3-13A. (All diodes are 1N4004 or equivalent.)

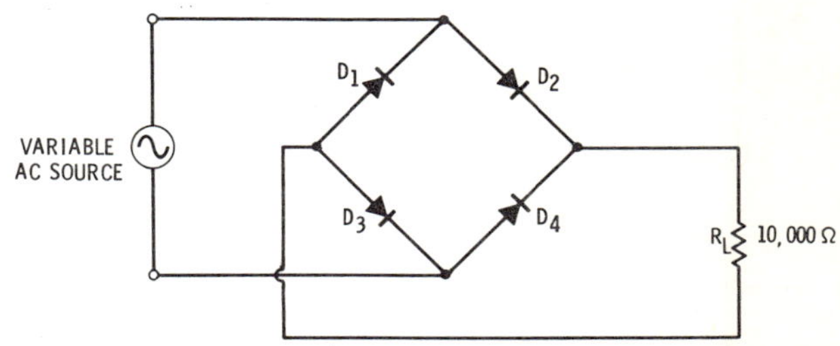

Fig. 3-13A. Single-phase, full-wave bridge rectifier circuit.

2. With a voltmeter, measure the ac voltage values across each diode and R_L, using 10 volts ac as the applied voltage. Record these values in Table 3-13A.
3. Adjust the ac power source to 20 volts ac and measure the ac voltage across each diode and across R_L. Record your values in Table 3-13A.
4. Repeat the same measurements with 30 volts ac applied and record in Table 3-13A.
5. Calculate the peak values of V_{in} and V_{R_L} and record in Table 3-13A.
6. Prepare the meter to measure dc voltage and complete the dc voltage values in Table 3-13B for 10, 20, and 30 volts ac applied.
7. Apply 10 volts ac to the bridge rectifier circuit. With an oscilloscope properly synchronized with the ac input, observe and record the necessary waveforms in Table 3-13C.

Table 3-13A. AC Values

Applied Voltage	V_{in} (peak)	V_{D_1}	V_{D_2}	V_{D_3}	V_{D_4}	V_{R_L} (rms)	V_{R_L} (peak)
10 V ac							
20 V ac							
30 V ac							

Table 3-13B. DC Values

Applied Voltage	V_{D_1}	V_{D_2}	V_{D_3}	V_{D_4}	V_{R_L}
10 V ac					
20 V ac					
30 V ac					

Table 3-13C. Oscilloscope Waveforms

		V_{peak}	$V_{p\text{-}p}$
V_{in}	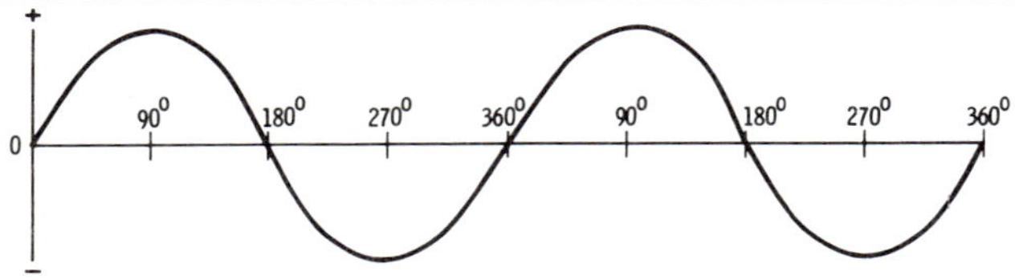		
V_{D_1}	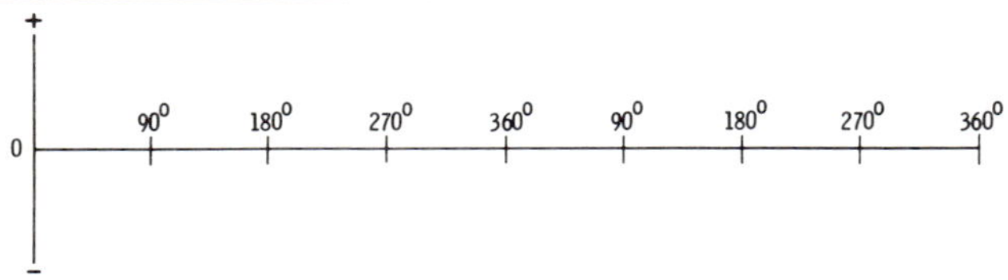		
V_{D_2}	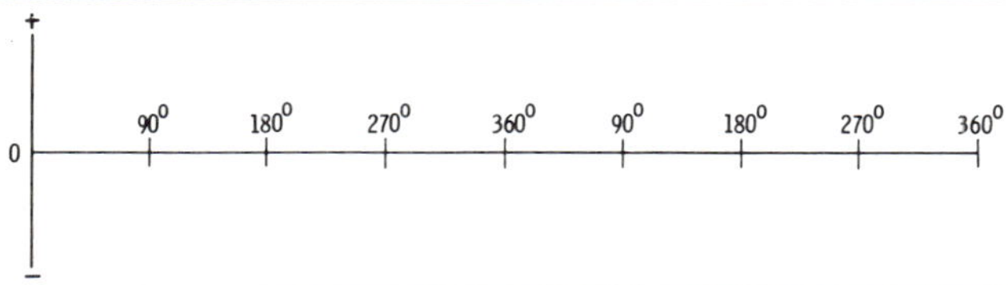		
V_{D_3}	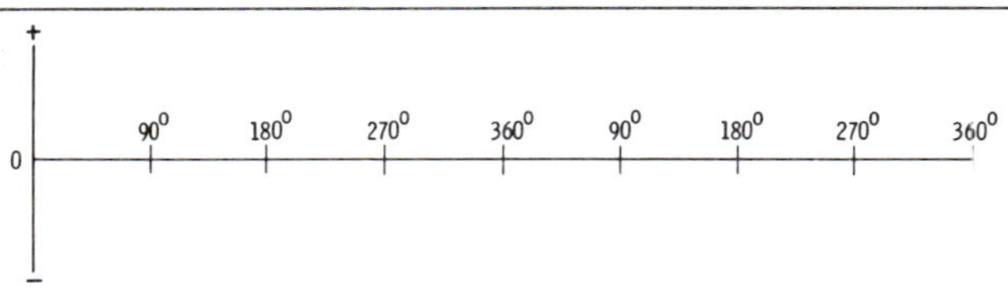		
V_{D_4}	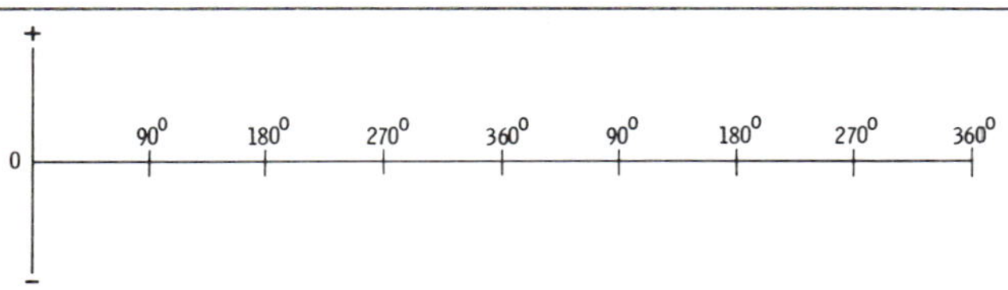		
V_{R_L}	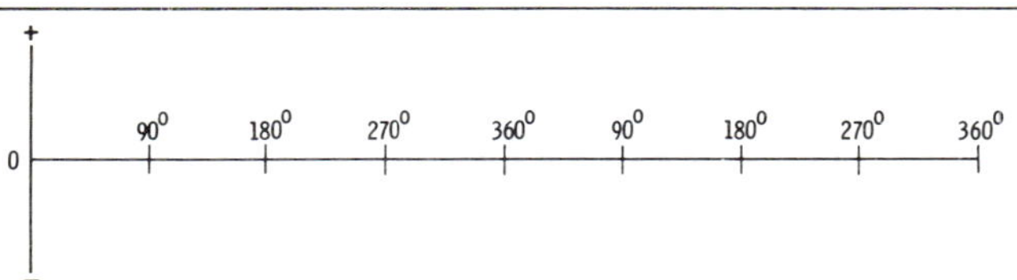		

8. Also, record the V_{peak} and $V_{p\text{-}p}$ values in Table 3-13C.
9. This concludes the experiment.

Analysis

1. What is the frequency of the ac ripple voltage for the bridge rectifier?

   ____

2. How do the dc output values with 10, 20, and 30 volts ac applied to the bridge circuit compare to those values obtained when using a single-phase, half-wave rectifier circuit?

   ____

   ____

   ____

3. What is the ratio of dc voltage output to the peak ac voltage input for this circuit?

$$\frac{V_{dc(out)}}{V_{peak(in)}} = \underline{\hspace{4cm}}.$$

4. How does the ratio obtained in Step 3 of the analysis compare to the full-wave rectifier that uses a center-tapped transformer?

5. Draw a diagram showing how vacuum-tube diodes could be used rather than semiconductor diodes to form a bridge rectifier circuit.

6. What would happen to the dc output if D_1 became short-circuited?

7. What would happen to the dc output if D_3 became an open-circuit path?

Power-Supply Filters—Capacitor Type

Objective

Rectification systems are used as the part of a power supply that converts an alternating-current input to a direct-current output. The output taken directly at the output of the rectifier circuit is a crude form of direct current which still contains a large amount of alternating current variation or ripple.

Several different types of filtering systems have been developed which can be added onto the output of a rectification system. The purpose of filtering systems is to modify the pulsating dc voltage at the output of a rectifier circuit so that the resulting voltage will be a purer or "smoother" form of direct current.

In this laboratory activity, you will see how a capacitor can be used to accomplish filtering. You will also observe some of the important power-supply design considerations, such as ripple factor, percentage ripple, and voltage regulation.

Equipment

Electronic multifunction meter.
Diodes (4)—1N4004 or equivalent.
Variable ac power supply.
Oscilloscope.
Resistors—15,000 Ω, 10,000 Ω, 5000 Ω, 1000 Ω, and 500 Ω.
Capacitor—40 μF, 100 V dcw.

Procedure

1. Construct the circuit shown in Fig. 3-14A.
2. Adjust the ac power supply to 10 volts (rms).

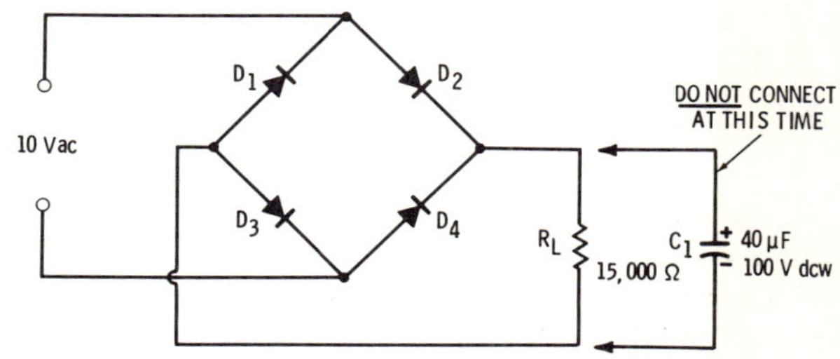

Fig. 3-14A. Capacitor filter.

3. With a properly calibrated oscilloscope, measure and record the output waveform (across R_L) in the following chart:

V_{R_L} Waveform	V_{peak}	$V_{p\text{-}p}$

4. Compute the following values for this rectification circuit:

(a) $V_{max} = 1.41 \times V_{rms} =$ _______ volts.

(b) $V_{dc} = 0.636 \times V_{max} =$ _______ volts.

(c) Ripple factor $= \dfrac{\text{ripple voltage (rms)}}{V_{dc}}$

$=$ _______.

5. Using the electronic multifunction meter, measure V_{max}, V_{dc}, and ripple voltage (rms). Record in Table 3-14A.

Table 3-14A. Power-Supply Values

V_{in} (rms)	V_{max} Computed	V_{max} Measured	V_{dc} Computed	V_{dc} Measured	Ripple Factor	Ripple %
10 V ac						
15 V ac						
20 V ac						
25 V ac						
30 V ac						

Table 3-14B. Power-Supply Resistive Loads

R_L	I_{R_L} Measured	V_R (rms) Measured	V_{dc} Computed*	V_{dc} Measured	Ripple Factor
15,000 Ω					
10,000 Ω					
5000 Ω					
1000 Ω					
500 Ω					

*V_{dc} for a circuit with a capacitor filter is calculated as:

$$V_{dc} = V_{max} - \left(\frac{V_{r(p\text{-}p)}}{2}\right)$$

6. Complete Table 3-14A for the indicated input voltages.
7. Turn off the power supply and connect a dc current meter in series with the load resistance. Adjust the power supply to 10 volts ac and measure the load current.

   Load current = ______ amperes dc.
8. Turn off the power supply and connect a 40-μF, 100-V dcw capacitor across R_L. Be sure to observe proper polarity.
9. Again, measure and record the load current with 10 volts ac applied.

   Load Current = ______ amperes dc.
10. With the oscilloscope, observe and record the waveform across R_L. Measure V_{dc} and ripple voltage (p-p) with the oscilloscope.

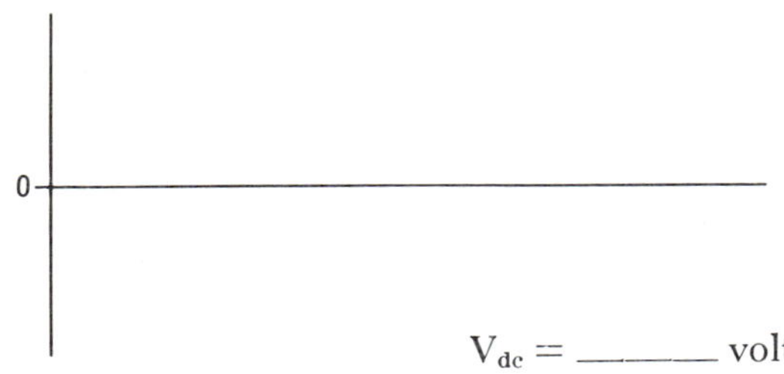

V_{dc} = ______ volts dc.

$V_{r(p\text{-}p)}$ = ______ volts ac.

11. Complete Table 3-14B for the indicated values of resistive loads with 10 volts ac applied.
12. This concludes the experiment.

Analysis

1. From Table 3-14A, when V_{in} was increased, what effect was observed on:
   (a) V_{max}? ______
   (b) V_{dc}? ______
   (c) Ripple factor? ______
2. What effect did the addition of a capacitor have on load current (see Steps 7 and 9)? Why?

   ______

   ______

   ______
3. What effect did the capacitor have on:
   (a) V_r(rms)? ______
   (b) V_{dc}? ______
   (c) Ripple factor? ______
4. From Table 3-14B, what effect does increased load have on:
   (a) V_r(rms)? ______
   (b) V_{dc}? ______
   (c) Ripple factor? ______
5. Calculate the percentage voltage regulation from Table 3-14B values, using 15,000 ohms as no-load and 500 ohms as full-load:

$$\%\ \text{Regulation} = \frac{V_{NL} - V_{FL}}{V_{FL}} \times 100 = \underline{\qquad}\ \%.$$

6. Calculate the ideal value of filter capacitor for the 5000-ohm load of Table 3-14B:

$$C = \frac{2.4 \times I_{dc}}{V_r(\text{rms})} = \underline{\qquad}\ \mu\text{F}.$$

Power-Supply Filters—RC and Pi (π) Type

Objective

Two improvements over the simple capacitor filter for dc power supplies are the RC and pi (π) filter networks. An RC filter section, consisting of one resistor and two capacitors, may be connected across the output of a rectifier circuit. The pi-(π) type filter is similar, except an inductor is used rather than a resistor. These types of filters will produce lower ripple voltage than the capacitor filter.

In this laboratory activity, you will be able to observe the operation of an RC filter section, and then a pi-(π) type filter connected onto the output of a bridge rectifier. You should note the improved filtering action of these two types of filters compared to the capacitor filter.

Equipment

Ac power source.
Electronic multifunction meter.
Resistors—1000 Ω, 470 Ω (2) and 270 Ω, 1.0 W.
Diodes (4)—1N4004 or equivalent.
Capacitors (2)—10 μF to 40 μF, with at least 50 V dcw rating.
Inductor—8 H to 16 H.

Procedure

Section A. RC Filter Circuit

1. Construct the RC filter circuit shown in Fig. 3-15A.
2. Apply 6.3 volts ac to the circuit. With a voltmeter, measure the ac voltage values across C_1, R_1, and R_L. Record these values in Table 3-15A.
3. Calculate the peak ac voltage values of V_{in}, V_{C_1}, V_{R_1}, and V_{R_L}.
4. Repeat the same measurements and calculations with 12.6 volts ac applied.
5. Prepare the meter to measure dc voltage and record the necessary dc values in Table 3-15B.

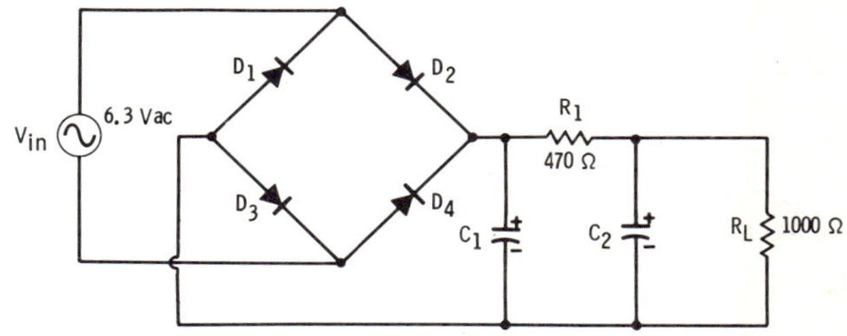

Fig. 3-15A. RC filter circuit.

Table 3-15B. DC Values (RC Filter)

V_{in}	V_{C_1}	V_{R_1}	V_{R_L}
6.3 V ac			
12.6 V ac			

Section B. Pi-(π) Type Filter Circuit

1. Replace R_1 with an inductor with a value in the range of 8 henrys to 16 henrys. Refer to this device as L_1. This makes the circuit a pi-(π) type filter.
2. Again, prepare the meter to measure ac voltage. Measure and record in Table 3-15C the values of V_{C_1}, V_{L_1}, and V_{R_L} with 6.3 volts ac applied.

Table 3-15A. AC Values (RC Filter)

V_{in}	V_{in} (peak)	V_{C_1}	V_{C_1} (peak)	V_{R_1}	V_{R_1} (peak)	V_{R_L}	V_{R_L} (peak)
6.3 V ac							
12.6 V ac							

Table 3-15C. AC Values—Pi-(π) Type Filter

V_{in}	V_{C_1}	V_{C_1} (peak)	V_{L_1}	V_{L_1} (peak)	V_{R_L}	V_{R_L} (peak)
6.3 V ac						
12.6 V ac						

3. Calculate the peak values of V_{C_1}, V_{L_1}, and V_{R_L} and record in Table 3-15C.
4. Repeat the same measurements and calculations with 12.6 volts ac applied. Record your values in Table 3-15C.
5. Prepare the meter to measure dc voltage. Measure and record the necessary dc values in Table 3-15D.

Table 3-15D. DC Values—Pi-(π) Type Filter

V_{in}	V_{C_1}	V_{L_1}	V_{R_L}
6.3 V ac			
12.6 V ac			

6. With 12.6 volts ac applied, substitute a 470-ohm, and then a 270-ohm resistor for the load resistor (R_L). Make the necessary measurements to complete Table 3-15E for the changes of load resistance.
7. Use the 1000-ohm resistor as the load (R_L) and maintain 12.6 volts ac as the source voltage.
8. With an oscilloscope properly synchronized with the ac input, record the necessary waveforms in Table 3-15F.
9. Also, record V_{peak} and $V_{p\text{-}p}$ values in Table 3-15F.
10. This concludes the experiment.

Table 3-15E. Load Variations—DC Values—Pi-(π) Type Filter

R_L	V_{C_1}	V_{L_1}	V_{R_L}
470 Ω			
270 Ω			

Analysis

1. From the data of Table 3-15B, compare V_{C_1} and V_{R_L} values. Why are these values different?

2. What characteristics of an RC filter does a comparison of the data of Tables 3-15A and B show?

Table 3-15F. Oscilloscope Waveforms—Pi-(π) Type Filter

		V_{peak}	$V_{p\text{-}p}$
V_{in}	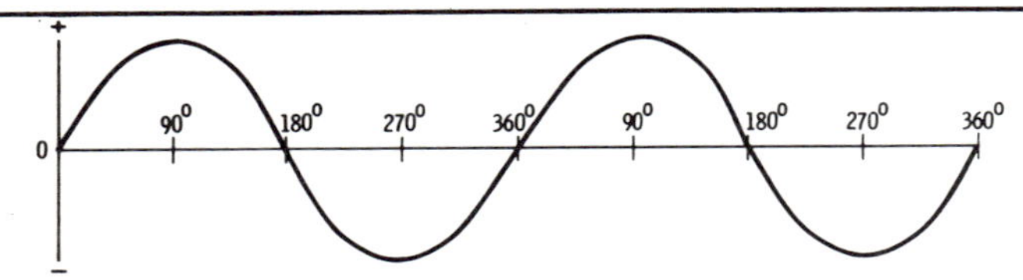		
V_{C_1}	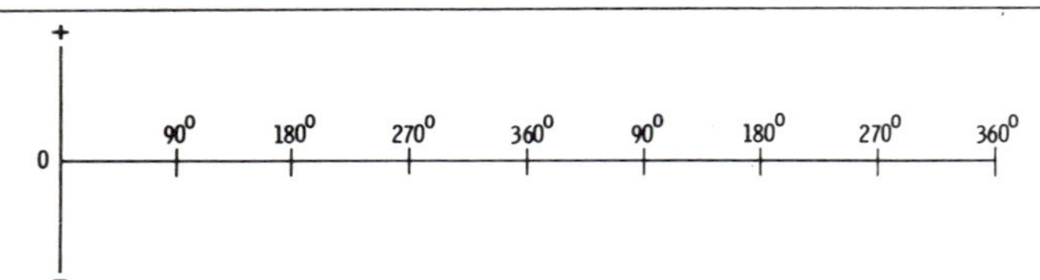		
V_{L_1}	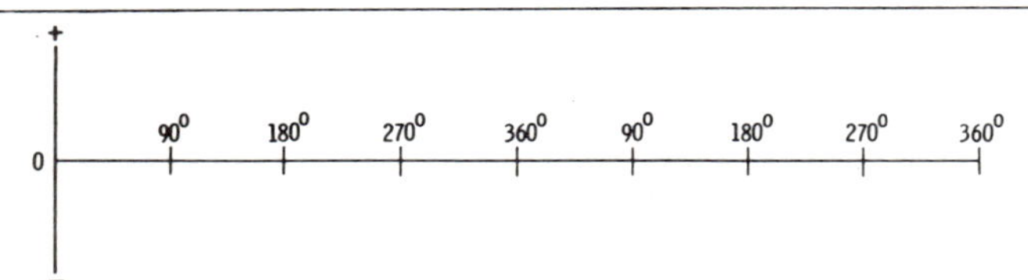		
V_{R_L}	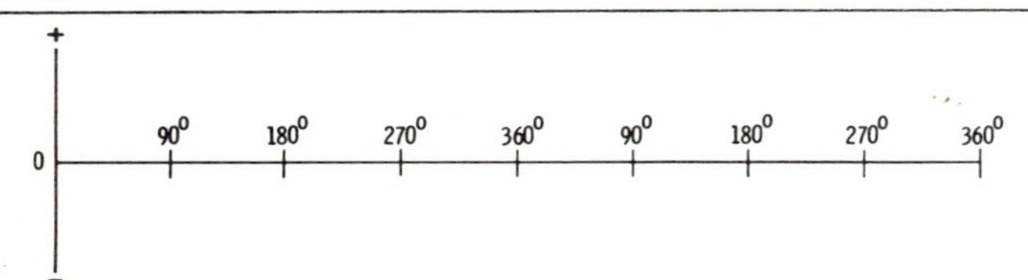		

3. For the RC filter, what is the ratio of dc voltage output to the peak ac voltage input?

$$\frac{V_{dc(out)}}{V_{peak(in)}} = \_\_\_\_\_\_\_\_\_\_\_\_.$$

4. What effect did replacing R_1 with an inductor have on the dc voltage output of the circuit? Compare Tables 3-15B and D.

5. From Tables 3-15B and D, how do the values of V_{R_1} and V_{L_1} compare? Why?

6. From the data of Table 3-15E, explain how a decrease in load resistance affects the dc output.

7. What effect would using a half-wave rectifier rather than a full-wave bridge rectifier have on the dc output from a pi (π) filter network?

8. Which type of filter (RC or π-type) is most desirable to use? Why?

9. For the π-type filter, what is the ratio of $V_{dc(out)}/V_{peak(in)}$ with a 1000-ohm load? With a 470-ohm load? With a 270-ohm load?

Half-Wave Voltage Doubler

Objective

In this laboratory activity you will observe the operation of a half-wave voltage doubler. A circuit such as this can be used to increase dc voltage levels from a fixed value ac source without using a transformer. In many cases, particularly where high voltages at low current levels are required, voltage-multiplier circuits are beneficial to use.

Equipment

Variable ac power supply.
Electronic multifunction meter.
Resistor—10,000 Ω.
Diodes (2)—1N4004 or equivalent.
Capacitors (2)—10 μF to 40 μF, with at least 50 V dcw rating.

Procedure

1. Construct the circuit shown in Fig. 3-16A.

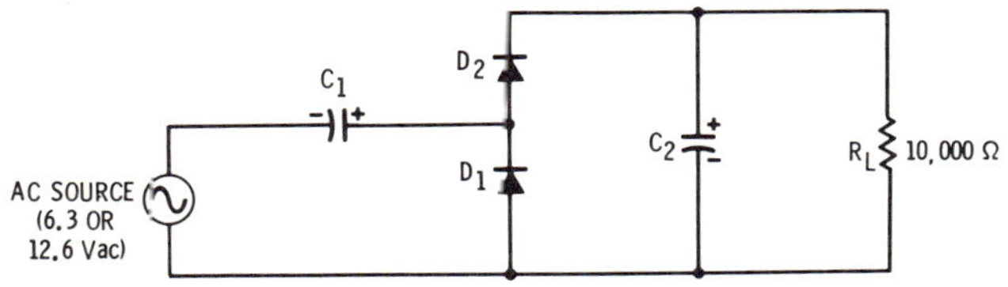

Fig. 3-16A. Half-wave, voltage-doubler circuit.

2. Apply 6.3 volts ac to the circuit. With a voltmeter, measure the ac voltage values across the diodes, capacitors, and R_L. Record these values in Table 3-16A.
3. Calculate the peak values of V_{in}, V_{C_1}, V_{C_2} and V_{R_L}. Record these values in Table 3-16A.
4. Repeat the same measurements and calculations with 12.6 volts ac applied.
5. Prepare the meter to measure dc voltage and complete the necessary dc values in Table 3-16B.

Table 3-16B. DC Values

V_{in}	V_{D_1}	V_{C_1}	V_{R_L}
6.3 V ac			
12.6 V ac			

6. With an oscilloscope properly synchronized with the ac input, record the necessary waveforms in Table 3-16C. Use 12.6 volts ac as the applied voltage.
7. Also, record V_{peak} and $V_{p\text{-}p}$ values in Table 3-16C.
8. This concludes the experiment.

Analysis

1. How is voltage multiplication accomplished?

___

___

___

2. How do the dc values of V_{C_1} and V_{R_L} compare in this circuit? Why?

___

___

___

3. What effect would a lower value resistor for R_L have on circuit operation?

___

___

___

Table 3-16A. AC Values

V_{in}	V_{in} (peak)	V_{D_1}	V_{D_2}	V_{C_1}	V_{C_1} (peak)	V_{C_2}	V_{C_2} (peak)	V_{R_L}	V_{R_L} (peak)
6.3 V ac									
12.6 V ac									

Table 3-16C. Oscilloscope Waveforms

		V_{peak}	$V_{p\text{-}p}$
V_{in}	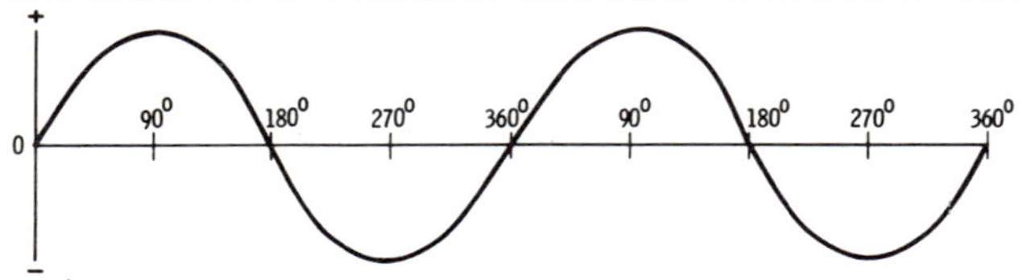		
V_{D_1}	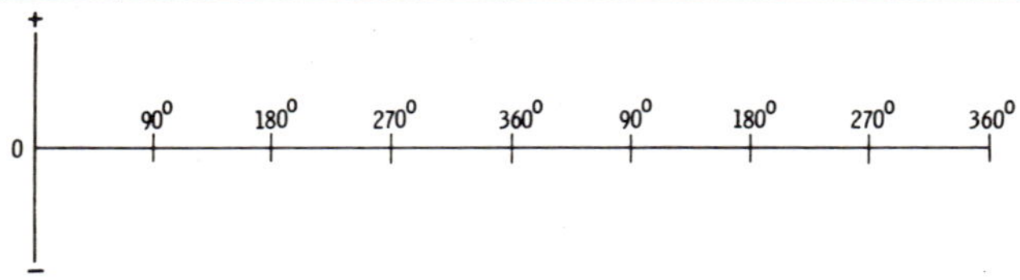		
V_{D_2}	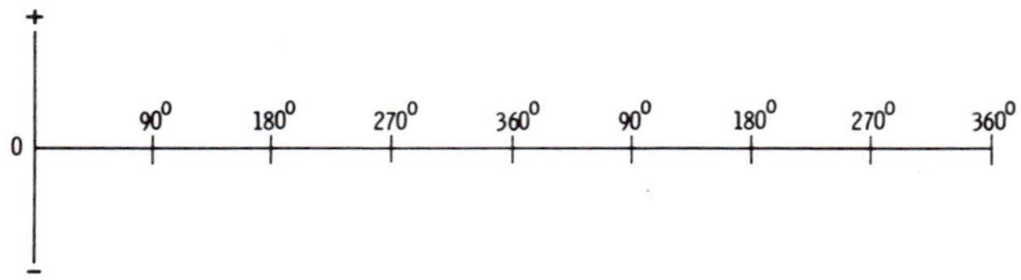		
V_{C_1}	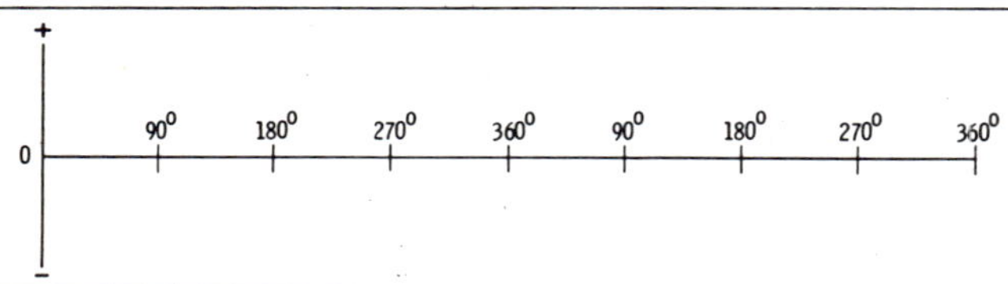		
V_{R_L}	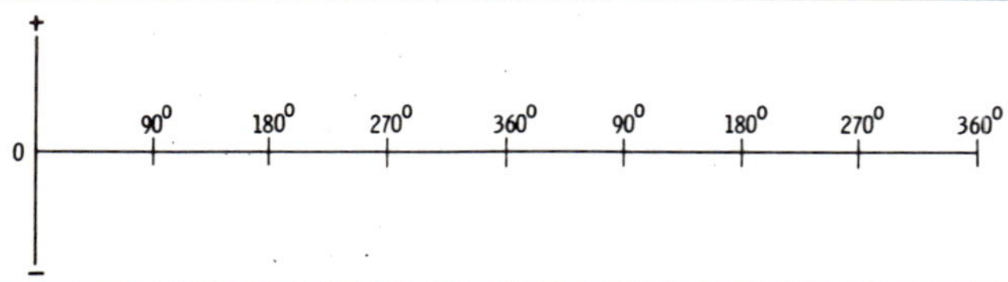		

4. What would happen if D_2 was reversed in this circuit?

____

____

____

5. What is the ratio of dc voltage output to peak ac voltage input for this circuit?

$$\frac{V_{dc(out)}}{V_{peak(in)}} = \_\_\_\_\_\_\_\_\_\_\_\_.$$

3-17

Full-Wave Voltage Doubler

Objective

In this laboratory activity, you will observe the operation of a full-wave voltage doubler. This circuit is similar in many respects to the half-wave voltage doubler. You should be able to observe some key differences between these two types of voltage doublers by comparing the data of this experiment with that of the previous experiment.

Equipment

Variable ac power supply.
Electronic multifunction meter.
Resistor—10,000 Ω.
Diodes (2)—1N4004 or equivalent.
Capacitors (2)—10 μF to 40 μF, with at least 50 V dcw rating.

Procedure

1. Construct the circuit shown in Fig. 3-17A.

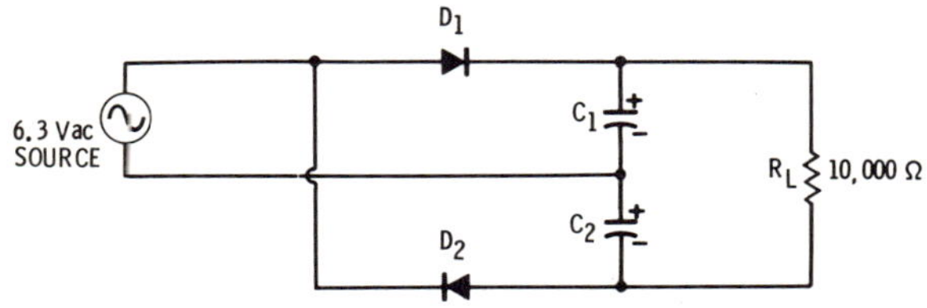

Fig. 3-17A. Full-wave, voltage-doubler circuit.

2. Apply 6.3 volts ac to the circuit. With a voltmeter, measure the ac voltage values across the diodes, capacitors, and R_L. Record these values in Table 3-17A.
3. Calculate the peak values of V_{in} and V_{R_L}. Record these values in Table 3-17A.
4. Repeat the same measurements and calculations with 12.6 volts ac applied.
5. Prepare the meter to measure dc voltage and complete the necessary dc values in Table 3-17B.
6. With an oscilloscope properly synchronized with the ac input, record the necessary waveforms in Table 3-17C. Use 12.6 volts ac as the applied voltage.
7. Also, record V_{peak} and $V_{p\text{-}p}$ values in Table 3-17C.
8. This concludes the experiment.

Analysis

1. What is the difference between the half-wave and full-wave voltage doubler?

____

____

____

2. How do the dc values of V_{C_1} and V_{R_L} compare in this circuit? Why?

____

____

____

Table 3-17A. AC Values

V_{in}	V_{in} (peak)	V_{D_1}	V_{D_2}	V_{C_1}	V_{C_2}	V_{R_L}	V_{R_L} (peak)
6.3 V ac							
12.6 V ac							

Table 3-17B. DC Values

V_{in}	V_{D_1}	V_{D_2}	V_{C_1}	V_{C_2}	V_{R_L}
6.3 V ac					
12.6 V ac					

3. What effect would a lower value resistor for R_L have on circuit operation?

____

____

____

4. What would happen if D_2 was reversed in this circuit?

____

____

____

5. What is the ratio of dc voltage output to peak ac voltage input for this circuit?

$$\frac{V_{dc(out)}}{V_{peak(in)}} = \_\_\_\_\_\_\_\_\_\_.$$

6. Compare the oscilloscope waveforms of Table 3-17C of this activity to Table 3-16C of the half-wave rectifier circuit.

____

____

____

Table 3-17C. Oscilloscope Waveforms

		V_{peak}	V_{p-p}
V_{in}	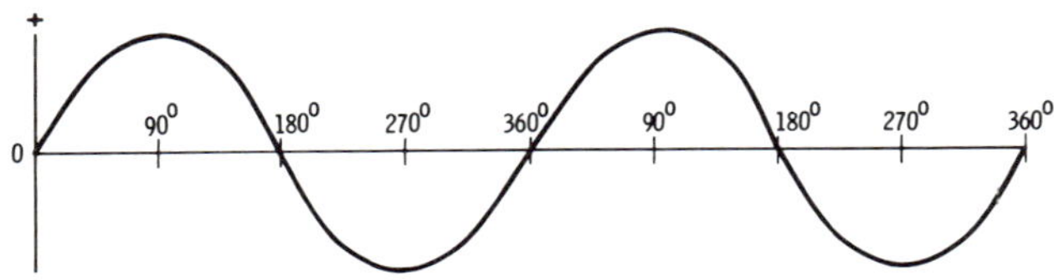 + 0 − 90° 180° 270° 360° 90° 180° 270° 360°		
V_{D_1}	+ 0 − 90° 180° 270° 360° 90° 180° 270° 360°		
V_{D_2}	+ 0 − 90° 180° 270° 360° 90° 180° 270° 360°		
V_{C_1}	+ 0 − 90° 180° 270° 360° 90° 180° 270° 360°		
V_{C_2}	+ 0 − 90° 180° 270° 360° 90° 180° 270° 360°		
V_{R_L}	+ 0 − 90° 180° 270° 360° 90° 180° 270° 360°		

3-18

Zener-Diode Voltage Regulation

Objective

An important characteristic of dc power sources is their ability to maintain a constant voltage output as load values change. This characteristic is referred to as the *voltage regulation* of a power source. In order to improve voltage regulation, a zener diode may be connected across the output of a filtered dc power source. The reverse-bias operational characteristics of this device make it desirable to maintain a constant voltage in a circuit.

In this laboratory activity, you will analyze the characteristics of a zener diode and then its operation as part of a dc power-supply circuit.

Equipment

Variable dc power supply.
Ac power source.
Diodes (4)—1N4004 or equivalent.
Electronic multifunction meter.
Zener diode—1N1775 or equivalent.
Capacitors—10 μF to 50 μF with 100 V dcw rating.
Inductor—8 H to 16 H.
Resistors—270 Ω, 500 Ω, 1000 Ω, 5000 Ω, and 15,000 Ω.

Procedure

Section A

In this section, you will observe the characteristics of a zener diode during forward- and reverse-bias operation. Zener diodes are used as voltage regulators due to their reverse-bias characteristics.

1. Construct the circuit shown in Fig. 3-18A.

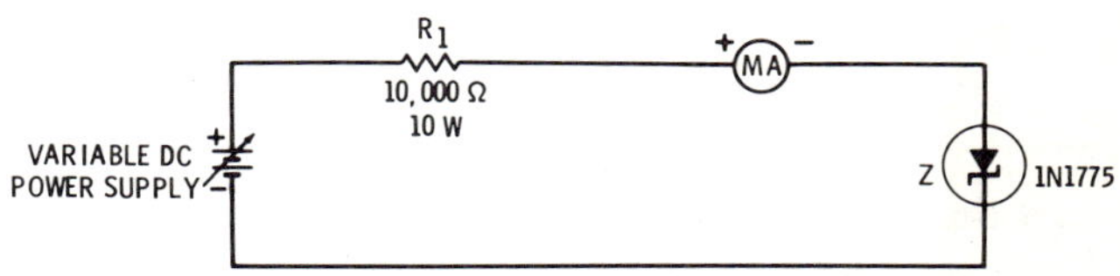

Fig. 3-18A. Circuit for observing zener diode forward- and reverse-bias characteristics.

2. Adjust the dc power supply from zero to 40 volts in the increments listed in Table 3-18A. Monitor the current in the circuit (I). Measure the voltage across the zener diode (V_Z) and the voltage across R_1 (V_{R_1}) for each increment of applied voltage. Record these values in Table 3-18A.
3. Turn off the power supply. Reverse the zener-diode terminals (anode to negative and cathode to positive). This will reverse-bias the zener diode.
4. Adjust the power supply from zero to 40 volts dc as in Step 2. Complete the necessary values in Table 3-18B.
5. Slowly increase the dc voltage from zero to observe the point where a current flow is first indicated on the meter. This is the reverse breakover voltage of the zener diode. Record this voltage value.

   Breakover voltage = ______ volts dc.
6. This concludes Section A of the activity.

Table 3-18A. Forward-Bias Characteristics

Applied Voltage	V_Z	V_{R_1}	I	$R_Z = V_Z/I$
0				
4				
8				
12				
16				
20				
30				
40				

Table 3-18B. Reverse-Bias Characteristics

Applied Voltage	V_Z	V_{R_1}	I	R_Z
0				
4				
8				
12				
16				
20				
30				
40				

Section B

In this part of the laboratory activity, you will use the zener diode as a voltage regulating device in a power-supply circuit. Observe its operation in relation to the characteristics observed in Section A.

1. Construct the circuit shown in Fig. 3-18B.

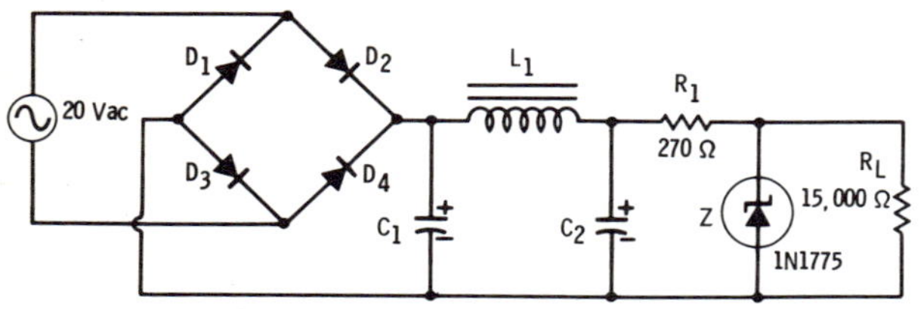

Fig. 3-18B. Circuit illustrating the use of a zener diode as a voltage regulating device.

2. Apply 20 volts ac to the input of this zener-diode-regulated power supply. Measure voltages V_{C_1}, V_{L_1}, V_{C_2}, V_{R_1}, and V_Z. Use these values to callate the dc current through L_1, R_1, Z, and R_L. Record the data in Table 3-18C.
3. Change the load resistance value to 5000 ohms and then to 500 ohms and complete Table 3-18C for each of these load resistance values.
4. Gradually reduce the ac input voltage from 20 volts ac while observing the voltage across R_L. At what ac voltage value does regulation stop?

   Regulation stops at ______ volts ac.
5. This concludes the experiment.

Analysis

1. Draw a graph using the data from Tables 3-18A and B to show the forward- and reverse-bias characteristics of the 1N1775 zener diode.

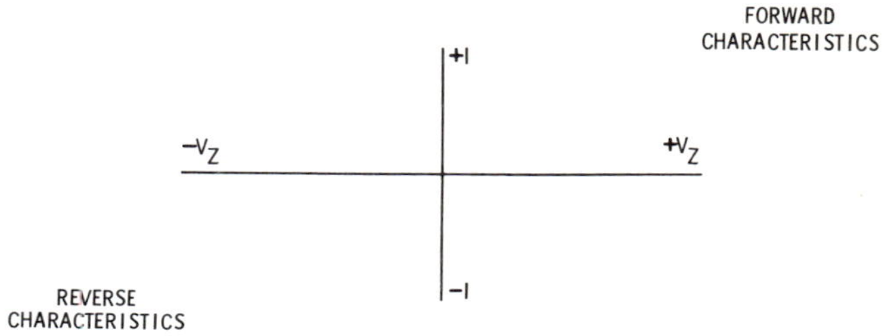

2. How does the zener diode differ from a conventional pn junction diode?

____

____

3. Why is the reverse-bias characteristic of a zener diode useful for voltage regulation?

____

____

4. How do the forward-bias characteristics of a zener diode compare with a pn junction diode?

____

____

Table 3-18C. Regulated Power-Supply Values

R_L	V_{C_1}	V_{L_1}	V_{C_2}	V_{R_1}	V_Z	$I_{L_1} = \frac{V_{L_1}}{X_{L_1}}$	$I_{R_1} = \frac{V_{R_1}}{R_1}$	$I_{R_L} = \frac{V_{R_L}}{R_L}$	$I_Z = I_{R_1} - I_{R_L}$
15,000 Ω									
5000 Ω									
500 Ω									

5. From the data of Table 3-18C, discuss the effect of a lower load resistance on:
   (a) V_{C_1}.
   (b) V_{C_2}.
   (c) V_{R_1}.
   (d) V_Z.
   (e) V_{R_L}.
   (f) I_{R_1}.
   (g) I_Z.
   (h) I_{R_L}.

6. Calculate the voltage regulation of this circuit using 15,000 ohms as no-load and 500 ohms as full-load.

$$\% \text{ Regulation} = \frac{V_{NL} - V_{FL}}{V_{FL}} = \_\_\_\_\_\_\_\_\_\_ \%.$$

7. What is the ratio of $V_{dc(out)}$ and $V_{peak(in)}$ for this circuit? (Use 15,000-ohm values.)

$$\frac{V_{dc(out)}}{V_{peak(in)}} = \_\_\_\_\_\_\_\_\_\_.$$

Gaseous-Tube Regulator

Objective

The ionization characteristics of a cold-cathode, gaseous-tube diode provide a simple method of voltage regulation. The cold-cathode tube has no heating element. The ionization voltage of a gas tube can be controlled by the type of gas used, the spacing of the elements, and the gas pressure inside the tube.

In this laboratory activity, you will use the OA2 cold-cathode diode as a voltage regulating device. You should compare the magnitude of the voltages associated with the gas-tube regulator with those of the zener regulator.

Equipment

Dc power source.
Cold-cathode, gaseous-tube diode—OA2.
Electronic multifunction meter.
Resistors—1000 Ω, 2500 Ω, 5000 Ω, and 10,000 Ω, 10 W.

Procedure

1. Construct the gaseous-tube diode, voltage regulator circuit shown in Fig. 3-19A.
2. Measure and record the following dc voltage and current values:

V_{R_1} = _______ volts.
V_{R_L} = _______ volts.
I_{R_1} = _______ amperes.
I_{V_1} = _______ amperes.
I_{R_L} = _______ amperes.

3. Increase the dc input voltage until I_{V_1} is equal to 28 mA. Measure and record the following values:

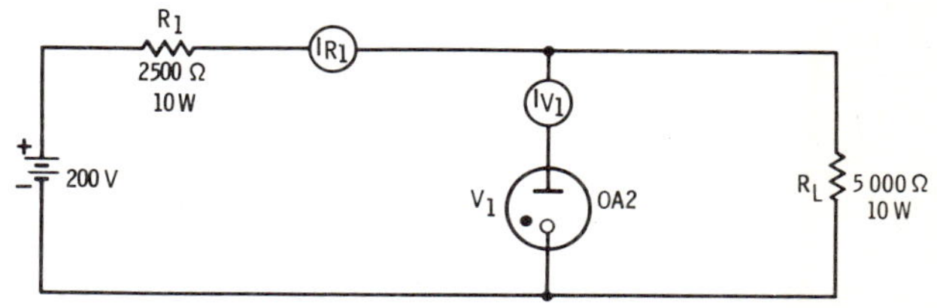

Fig. 3-19A. Gaseous-tube diode, voltage regulator circuit.

V_{source} = _______ volts.
V_{R_1} = _______ volts.
V_{R_L} = _______ volts.
I_{R_1} = _______ amperes.
I_{R_L} = _______ amperes.

4. Decrease the dc input voltage until I_{V_1} is equal to 4 mA. Measure and record the following values:

V_{source} = _______ volts.
V_{R_1} = _______ volts.
V_{R_L} = _______ volts.
I_{R_1} = _______ amperes.
I_{R_L} = _______ amperes.

5. Turn off the dc power source and replace R_L with a 10,000-ohm resistor. Adjust the dc input voltage to 200 volts and complete the measurements indicated in Table 3-19A.
6. Transfer the data from Step 3 to the 5000-ohm line of Table 3-19A.
7. Turn off the power supply and replace R_L with a 2500-ohm resistor. Maintain the dc input at 200 volts. Make the necessary measurements to complete the 2500-ohm line of Table 3-19A.
8. This completes the experiment.

Table 3-19A. Voltage Regulation Values

R_L	V_{R_1}	V_{R_L}	I_{R_1}	I_{V_1}	I_{R_L}
10,000 Ω					
5000 Ω					
2500 Ω					

Analysis

1. How does the cold-cathode gaseous tube accomplish voltage regulation?

2. From the data of Step 2, verify that

$$I_{R_1} = I_{V_1} + I_{R_L}.$$

3. From Step 3, where I_{V_1} is relatively high, how do the following values vary from the previous step:

I_{R_1}?

I_{R_L}?

V_{R_1}?

V_{R_L}?

4. From Step 4, where I_{V_1} is relatively low, how do the following values vary from Step 2:

I_{R_1}?

I_{R_L}?

V_{R_1}?

V_{R_L}?

5. From the data of Table 3-19A, what effect does a decrease in load resistance have on:

I_{R_1}?

I_{R_L}?

V_{R_1}?

V_{R_L}?

6. Calculate the voltage regulation of the circuit:*

$$\% \text{ Regulation} = \frac{V_{NL} - V_{FL}}{V_{FL}} \times 100 = \_\_\_\_\_\_ \%.$$

*Use V_{NL} as the 10,000-ohm load and V_{FL} as the 2500-ohm load.

3-20

Series Transistor Voltage Regulator

Objective

The purpose of a voltage regulator circuit of a dc power supply is to maintain a constant value of dc voltage over a range of loads. In this laboratory activity, you will analyze the operation of a series transistor voltage regulator.

Equipment

Variable dc source.
Transistors (2)—2N1184.
Potentiometer—10,000 Ω.
Resistors—10,000 Ω (2), 1000 Ω (2), 560 Ω, 500 Ω, 100 Ω (2).
Dry cell—1.5 V.
Electronic multifunction meter.

Procedure

1. Construct the series transistor, voltage regulator circuit shown in Fig. 3-20A.

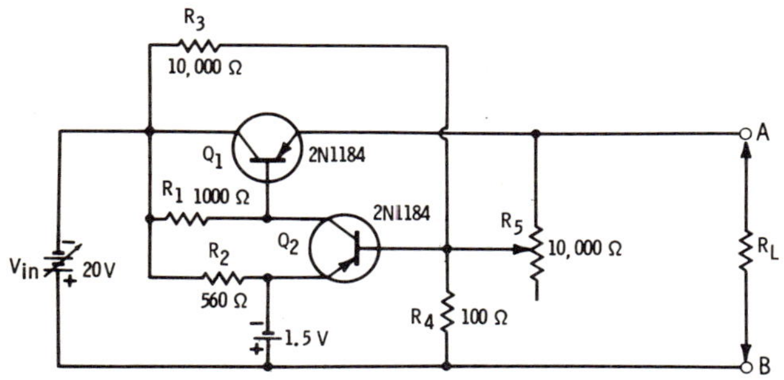

Fig. 3-20A. Series transistor, voltage regulator circuit.

2. Adjust the input voltage to the regulator circuit to 20 volts dc, and adjust R_5 for 10 volts output across points A and B with no R_L connected.
3. Change the value of the load resistor to 10,000, 1000, 500, and 100 ohms while observing the regulation characteristic of the circuit. Record the output voltage associated with each load in Table 3-20A.
4. Alter the input voltage applied to the circuit and determine the range of regulation. Do not exceed 30 volts dc input.

   Range of voltage regulation = ______ volts dc to ______ volts dc.

5. This concludes the experiment.

Table 3-20A. Output Voltage Values

R_L	Output Voltage
∞	10 V dc
10,000 Ω	
1000 Ω	
500 Ω	
100 Ω	

Analysis

1. Determine the voltage regulation of the circuit using no-load voltage equal to 10 volts dc and full-load voltage equal to the value measured at 1000-ohm load:

$$\% \text{ Regulation} = \frac{V_{NL} - V_{FL}}{V_{FL}} \times 100 = \_\_\_\_\_\_ \%.$$

2. Determine the regulation of the circuit using $V_{NL} = 10$ volts dc and V_{FL} equal to the value measured at 100-ohm load.

$$\% \text{ Regulation} = \frac{V_{NL} - V_{FL}}{V_{FL}} \times 100 = \_\_\_\_\_\_ \%.$$

3. Analyze the operation of this voltage regulator circuit, pointing out the function of Q_1, Q_2, R_3, R_5, and the 1.5-volt dry cell.

____

____

____

____

Shunt Transistor Voltage Regulator

Objective

In this laboratory activity, you will observe the operation of a shunt transistor, voltage regulator circuit. This circuit is used to maintain a constant dc voltage output from a dc source such as a bridge rectifier.

Equipment

Variable dc source.
Transistor—2N1184.
Potentiometer—10,000 Ω.
Dry cell—1.5 V.
Resistors—10,000 Ω, 5000 Ω, 1000 Ω, and 270 Ω.
Electronic multifunction meter.

Procedure

1. Construct the shunt transistor, voltage regulator circuit shown in Fig. 3-21A.

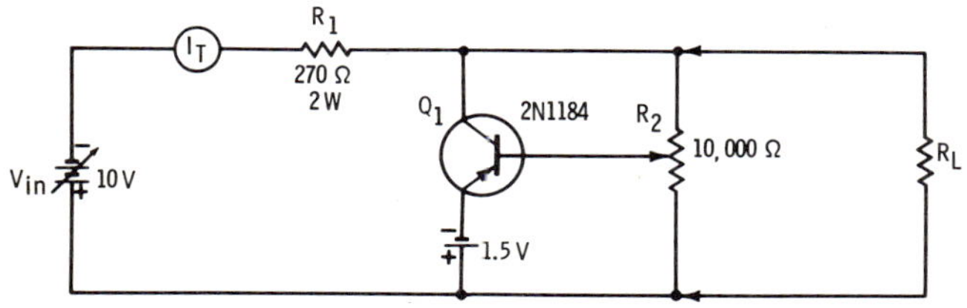

Fig. 3-21A. Shunt transistor, voltage regulator circuit.

2. Determine the output voltage, total current (I_T), collector current (I_C), base current (I_B), and the load current (I_L) for each load value indicated in Table 3-21A. Adjust R_2 to 5 volts output with no load connected.
3. Reduce the value of dc input voltage (V_{in}) in 5-volt increments starting with 20 volts, and change from a 10,000-ohm to a 1000-ohm load for each increment. Measure these output values and record them in Table 3-21B.
4. This concludes the experiment.

Table 3-21B. Input/Output Voltage Values

V_{in}	V_{R_L} (10,000 Ω)	V_{R_L} (1000 Ω)
20		
15		
10		
5		

Analysis

1. Determine the regulation from 10,000 ohms (no load) to 1000 ohms (full-load) for 5-volts dc output from the data of Table 3-21A.

$$\% \text{ Regulation} = \frac{V_{NL} - V_{FL}}{V_{FL}} \times 100 = \_\_\_\_\_\_ \%.$$

2. From the data of Table 3-21B, where does voltage regulation at low voltages drop out?

____

____

____

3. Explain the characteristics and limitations of a shunt transistor regulator.

____

____

____

Table 3-21A. Shunt Regulator Values

R_L	V_{R_L}	I_T	I_C	I_B	I_L
∞					
10,000 Ω					
5000 Ω					
1000 Ω					

4. How could a more sensitive regulation circuit be achieved? Draw a circuit showing how this could be accomplished.

5. Write an analysis of the operation of this circuit, pointing out the functions of Q_1, R_1, R_2, and the 1.5-volt dry cell.

____

____

____

____

3-22

Transistor Current Regulator

Objective

In this laboratory activity, you will construct and analyze a simple transistor circuit that could be used as a current limiter in a dc power supply. A current limiter maintains a constant level of current through the load connected to the power source.

Equipment

Variable dc source.
Transistor—2N1184.
Dry cell—1.5 V.
Resistors—10,000 Ω, 4700 Ω, 1000 Ω (2 W), 270 Ω, 100 Ω, and 10 Ω.
Electronic multifunction meter.

Procedure

1. Construct the transistor current-regulator circuit illustrated in Fig. 3-22A.

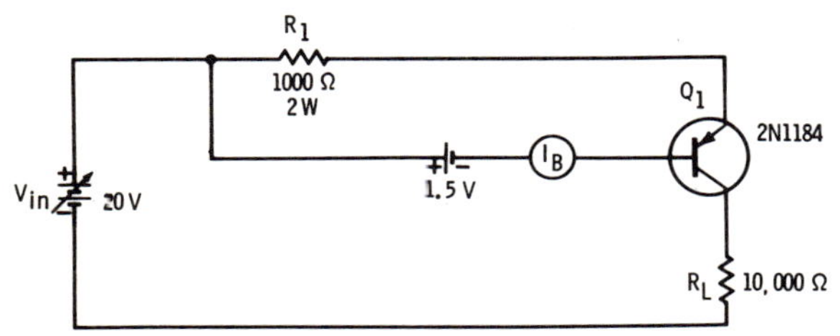

Fig. 3-22A. Transistor current-regulator circuit.

2. Measure I_T (total current), I_B (base current), I_L (load current) and V_{R_L} for each value of R_L. Record these values in Table 3-22A.
3. This concludes the experiment.

Analysis

1. Compare the values of I_L (load current) and I_B (base current) for each value of R_L.

____

____

____

2. Determine the current regulation from 10,000 ohms (no load) to 10 ohms (full load).

$$\% \text{ Regulation} = \frac{I_{FL} - I_{NL}}{I_{FL}} \times 100 = \_\_\_\_\_\_ \%.$$

3. Explain how this circuit could be improved to achieve better regulation.

____

____

____

4. Write an analysis of the operation of this circuit, pointing out the function of each component.

____

____

____

____

Table 3-22A. Current Regulator Values

R_L	I_T	I_B	I_L	V_{R_L}
10,000 Ω				
4700 Ω				
270 Ω				
100 Ω				
10 Ω				

Three-Phase, Half-Wave Rectification

Objective

Large quantities of direct current used in industry are most economically supplied by three-phase rectification systems. Three-phase rectifiers produce an output that has less ac ripple content than single-phase rectifiers have.

In this laboratory activity, you will observe the operation of a three-phase, half-wave rectification system. You will use three semiconductor diodes as the rectifying devices. It would also be possible to use three vacuum tubes or gaseous tubes as the rectifiers.

Equipment

Multifunction meter.
Variable, three-phase ac power source.
Diodes (3)—1N4004 or equivalent.
Single-pole, single-throw switches (3).
Resistor—10,000 Ω, 10 W.
Oscilloscope.

Procedure

1. Construct the three-phase, half-wave rectifier circuit shown in Fig. 3-23A. A three-phase, four-wire system is needed to provide a neutral.
2. Adjust the power source to 10-volts ac input per phase to the rectifier circuit. The three switches should be open.
3. Measure the dc voltage across R_L during the following conditions:
   (a) Close switch in phase A.

   V_{R_L} = ________ volts dc.

   (b) Close switches in phases A and B.

   V_{R_L} = ________ volts dc.

   (c) Close switches in phases A, B, and C.

   V_{R_L} = ________ volts dc.
4. With the switches closed, perform the necessary ac voltage readings and calculations to complete Table 3-23A.
5. Adjust the power source to 20 volts ac, then 30 volts ac in order to complete Table 3-23A.

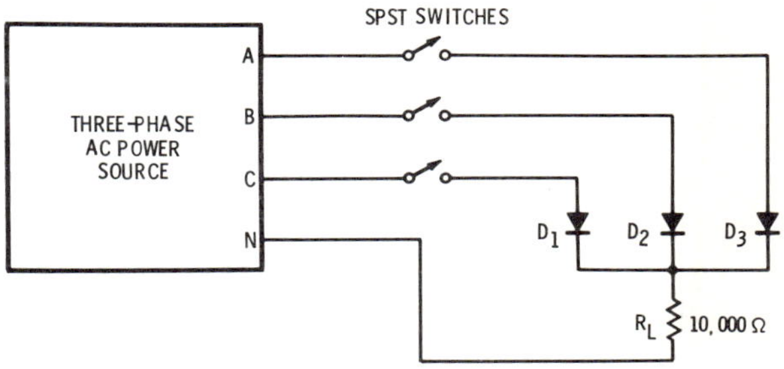

Fig. 3-23A. Three-phase, half-wave rectifier circuit.

6. Prepare the meter to measure dc voltage and complete the dc voltage values in Table 3-23B.

Table 3-23A. AC Values

Applied Voltage	V_{D_1}	V_{D_2}	V_{D_3}	V_{R_L}
10 V ac				
20 V ac				
30 V ac				

7. Apply 10 volts ac to the input. With an oscilloscope properly synchronized with phase A of the ac input, observe and record the necessary waveforms in Table 3-23C. Also, record the V_{peak} and $V_{p\text{-}p}$ values.
8. Open each of the spst switches. The ac input should remain at 10 volts ac.

Table 3-23B. DC Values

Applied Voltage	V_{D_1}	V_{D_2}	V_{D_3}	V_{R_L}
10 V ac				
20 V ac				
30 V ac				

9. Record the following oscilloscope displays of the voltage across the load during each of the conditions listed in Table 3-23D. Also record the V_{peak} and $V_{p\text{-}p}$ values.
10. This concludes the experiment.

Table 3-23C. Oscilloscope Waveforms

		V_{peak}	$V_{p\text{-}p}$
$V_{A\text{-}N}$	+ 0 − 90° 180° 270° 360° 90° 180° 270° 360°		
$V_{B\text{-}N}$	+ 0 − 90° 180° 270° 360° 90° 180° 270° 360°		
$V_{C\text{-}N}$	+ 0 − 90° 180° 270° 360° 90° 180° 270° 360°		
V_{R_L}	+ 0 − 90° 180° 270° 360° 90° 180° 270° 360°		

Table 3-23D. Output Waveforms

Circuit Condition	Load Voltage Waveform	V_{peak}	$V_{p\text{-}p}$
Switch in Phase A Closed	+ 0 − 90° 180° 270° 360° 90° 180° 270° 360°		
Switches in Phases A and B Closed	+ 0 − 90° 180° 270° 360° 90° 180° 270° 360°		
Switches in Phases A, B, and C Closed	+ 0 − 90° 180° 270° 360° 90° 180° 270° 360°		

Analysis

1. What is the frequency of the ac ripple across R_L?

____

2. What is the ratio of the dc voltage output to the peak ac voltage input for this rectifier circuit?

$$\frac{V_{dc(out)}}{V_{peak(in)}} = \_\_\_\_\_\_\_\_\_\_.$$

3. Draw a diagram showing how three vacuum-tube diodes could be used in place of the semiconductor diodes to form a three-phase, half-wave rectifier.

4. What effect does increasing the applied ac voltage have on the dc output of this circuit?

5. What would happen to the output if the polarity of D_3 was reversed?

6. What would happen to the output if D_2 became shorted?

7. In Step 3, what occurred as the switches were successively closed?

8. What is the phase relationship between the top three waveforms in Table 3-23C?

9. What do the waveforms of Table 3-23D illustrate?

10. Why is three-phase rectification more desirable for most industrial applications than single-phase rectification?

3-24

Three-Phase, Full-Wave Rectification

Objective

Three-phase, full-wave rectifier circuits produce a dc output that has less ac ripple than their half-wave counterparts have. In addition, they may be used with either a wye- or delta-connected, three-phase system. However, six diodes are required for the operation of a three-phase, full-wave rectifier.

In this laboratory activity, you will observe the operation of a three-phase, full-wave rectification system using six semiconductor diodes.

Equipment

Multifunction meter.
Variable three-phase power source.
Diodes (6)—1N4004 or equivalent.
Spst switches (3).
Resistor—10,000 Ω, 10 W.
Oscilloscope.

Procedure

1. Construct the three-phase, full-wave rectifier circuit shown in Fig. 3-24A.

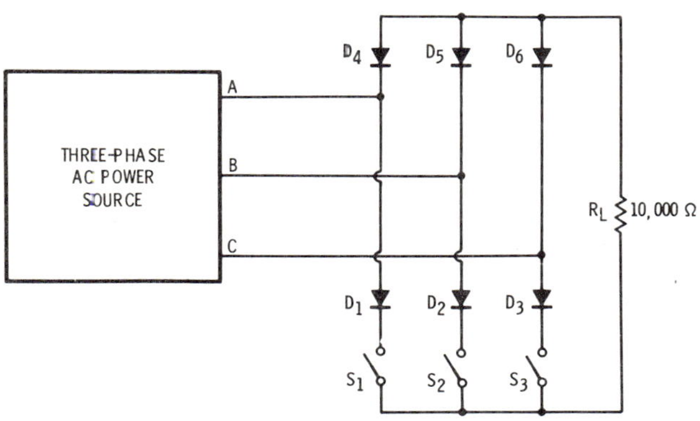

Fig. 3-24A. Three-phase, full-wave rectifier circuit.

2. Adjust the power source to 10 volts ac. The three switches should be open.
3. Measure the dc voltage across R_L during the following conditions:
   (a) All three switches open.

   V_{R_L} = ______ volts dc.

   (b) Switch S_1 closed.

   V_{R_L} = ______ volts dc.

   (c) Switches S_1 and S_2 closed.

   V_{R_L} = ______ volts dc.

   (d) All three switches closed.

   V_{R_L} = ______ volts dc.
4. With all three switches closed, perform the necessary ac voltage readings and calculations to complete Table 3-24A.
5. Adjust the power source to 20 volts ac, then 30 volts ac in order to complete Table 3-24A.
6. Prepare the meter to measure dc voltage and complete the dc voltage values in Table 3-24B.
7. Apply 10 volts ac to the input. With an oscilloscope properly synchronized with phase A of the ac input, observe and record the necessary waveforms in Table 3-24C. Also, record the V_{peak} and $V_{p\text{-}p}$ values.
8. Open all three switches, S_1, S_2, and S_3. The ac input should remain at 10 volts ac.
9. Record the oscilloscope displays of the voltage across the load during each of the conditions listed in Table 3-24D. Also, record the V_{peak} and $V_{p\text{-}p}$ values.
10. This concludes the experiment.

Analysis

1. What is the frequency of the ac ripple across R_L?

____

Table 3-24A. AC Values

Applied Voltage	V_{in} (peak)	V_{D_1}	V_{D_2}	V_{D_3}	V_{D_4}	V_{D_5}	V_{D_6}	V_{R_L}	V_{R_L} (peak)
10 V ac									
20 V ac									
30 V ac									

Table 3-24B. DC Voltage Values

Applied Voltage	V_{D_1}	V_{D_2}	V_{D_3}	V_{D_4}	V_{D_5}	V_{D_6}	V_{R_L}
10 V ac							
20 V ac							
30 V ac							

Table 3-24C. Oscilloscope Waveforms

		V_{peak}	V_{p-p}
V_{D_1}	+ 0 90° 180° 270° 360° 90° 180° 270° 360° −		
V_{D_2}	+ 0 90° 180° 270° 360° 90° 180° 270° 360° −		
V_{D_3}	+ 0 90° 180° 270° 360° 90° 180° 270° 360° −		
V_{D_4}	+ 0 90° 180° 270° 360° 90° 180° 270° 360° −		
V_{D_5}	+ 0 90° 180° 270° 360° 90° 180° 270° 360° −		
V_{D_6}	+ 0 90° 180° 270° 360° 90° 180° 270° 360° −		
V_{R_L}	+ 0 90° 180° 270° 360° 90° 180° 270° 360° −		

Table 3-24D. Output Waveforms

Circuit Condition	Load Voltage Waveform	V_{peak}	$V_{p\text{-}p}$
All Switches Open	+ 0 90° 180° 270° 360° 90° 180° 270° 360° −		
S_1 Closed	+ 0 90° 180° 270° 360° 90° 180° 270° 360° −		
S_1 and S_2 Closed	+ 0 90° 180° 270° 360° 90° 180° 270° 360° −		
S_1, S_2, and S_3 Closed	+ 0 90° 180° 270° 360° 90° 180° 270° 360° −		

2. What is the ratio of the dc voltage output to the peak ac voltage input for this rectifier circuit?

$$\frac{V_{dc(out)}}{V_{peak(in)}} = \_\_\_\_\_\_\_\_\_\_.$$

3. Compare the dc voltage output of this circuit to that of the three-phase, half-wave rectifier.

4. What would happen to the output if the polarity of D_4 was reversed?

5. What would happen to the output if D_5 became shorted?

6. In Step 3, what occurred when the switches were successively closed?

7. What are the phase relationships of the voltages across the diodes (see Table 3-24C)?

8. What do the waveforms of Table 3-24D illustrate?

Three-Phase Rectification Using SCRs

Objective

In previous laboratory activities, three-phase rectification was investigated. The basic principles of three-phase rectification may be employed by utilizing SCRs rather than pn junction diodes. When SCRs are used, it is possible to delay the firing time of any rectifier and thus control the output voltage and current flow through the load device.

In this laboratory activity, you will construct a three-phase, half-wave rectifier circuit that uses SCRs. Conduction through each three-phase line will be determined by the firing time of the SCR in that line. An RC phase-shift control network will be used to control the firing of the SCRs.

Equipment

Three-phase power source.
Potentiometers (3)—5000 Ω.
Resistors—1000 Ω (3); 10,000 Ω, 10 W.
Capacitors (3)—0.1 μF.
Silicon controlled rectifiers (3)—GE-C22B.
Multifunction meter.

Procedure

1. Construct the three-phase, SCR half-wave rectifier circuit shown in Fig. 3-25A.
2. Adjust the power source to 30 volts ac input to the circuit.
3. Connect a voltmeter across R_L and monitor the dc voltage as the SCRs are fired. None of the SCRs should be fired initially.
4. Adjust R_1 until SCR_1 conducts. Record the output across R_L.

   V_{R_L} = ______ volts dc.
5. Adjust R_2 until SCR_2 conducts. Record the output across R_L with both SCR_1 and SCR_2 conductive.

   V_{R_L} = ______ volts dc.
6. Adjust R_3 until SCR_3 conducts. Record the output with all three SCRs conductive.

   V_{R_L} = ______ volts dc.
7. With all SCRs conductive, perform the necessary ac voltage readings and calculations to complete Table 3-25A.

Table 3-25A. AC Values

Applied Voltage	V_{SCR_1}	V_{SCR_2}	V_{SCR_3}	V_{R_L}
30 V ac				
40 V ac				
50 V ac				

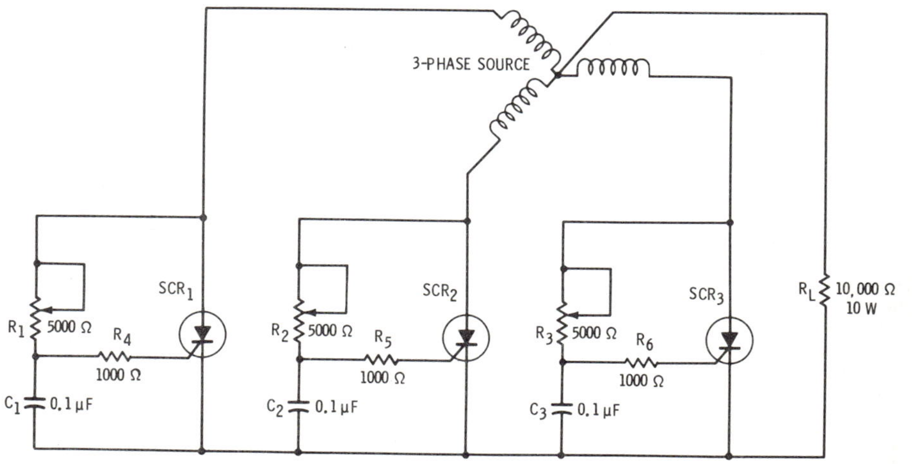

Fig. 3-25A. Three-phase, half-wave rectifier circuit using SCRs.

8. Adjust the power source to 40 volts ac, then 50 volts ac in order to complete Table 3-25A.
9. Prepare the meter to measure dc voltage and complete the dc voltage values in Table 3-25B.

Table 3-25B. DC Values

Applied Voltage	V_{SCR_1}	V_{SCR_2}	V_{SCR_3}	V_{R_L}
30 V ac				
40 V ac				
50 V ac				

10. Apply 30 volts ac to the input. With an oscilloscope properly synchronized with Phase A of the ac input, observe and record the necessary waveforms in Table 3-25C. Also, record the V_{peak} and $V_{p\text{-}p}$ values.
11. Adjust R_1, R_2, and R_3 until the SCRs no longer conduct. The ac input should remain at 30 volts ac.
12. Record the following oscilloscope displays of the voltage across the load during each of the conditions listed in Table 3-25D. Also, record the V_{peak} and $V_{p\text{-}p}$ values. This is the same procedure as Steps 4 through 6.
13. This concludes the experiment.

Table 3-25C. Oscilloscope Waveforms

		V_{peak}	$V_{p\text{-}p}$
$V_{A\text{-}N}$	+ 0 − 90° 180° 270° 360° 90° 180° 270° 360°		
$V_{B\text{-}N}$	+ 0 − 90° 180° 270° 360° 90° 180° 270° 360°		
$V_{C\text{-}N}$	+ 0 − 90° 180° 270° 360° 90° 180° 270° 360°		
V_{R_L}	+ 0 − 90° 180° 270° 360° 90° 180° 270° 360°		

Analysis

1. What is the frequency of the ac ripple across R_L?

____

2. What is the ratio of the dc voltage output to the peak ac voltage input for this circuit?

$$\frac{V_{dc(out)}}{V_{peak(in)}} = \_\_\_\_\_\_\_\_\_\_.$$

3. What effect does increasing the applied ac voltage have on the dc output of this circuit? (See Table 3-25B.)

____

____

____

Table 3-25D. Output Waveforms

Circuit Condition	Load Voltage Waveform	V_{peak}	$V_{p\text{-}p}$
SCR_1 Conductive	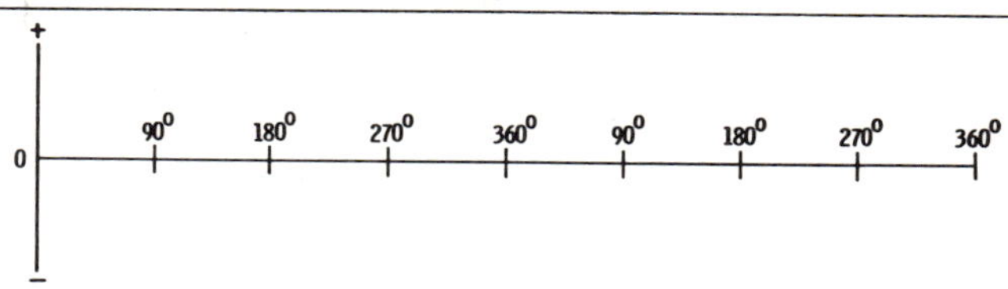		
SCR_1 and SCR_2 Conductive	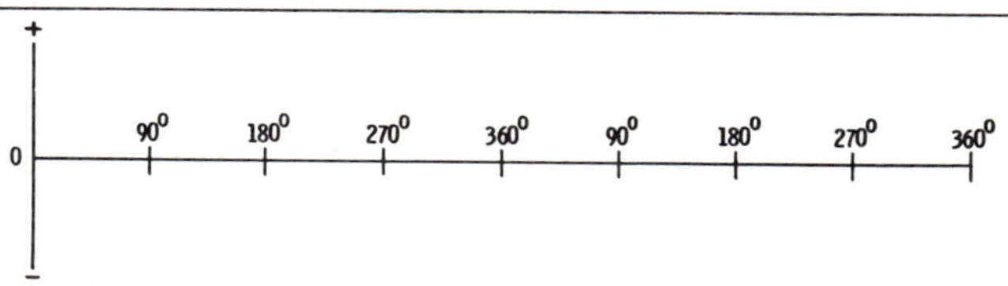		
All SCRs Conductive	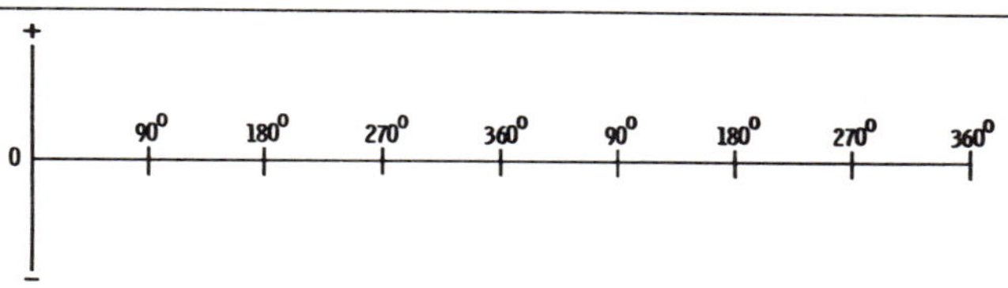		

4. In Steps 4 through 6, what occurred as the SCRs successively became conductive?

____

____

____

5. What do the waveforms of Table 3-25D illustrate?

____

____

____

6. Draw a diagram showing how three thyratrons could be used in place of the SCRs to form a three-phase, half-wave rectifier.

Integrated-Circuit Voltage Regulator

Objective

In this laboratory activity, you will use an integrated circuit (IC) to accomplish voltage regulation. This type of regulator circuit is the most recent development among voltage regulating devices. The miniature construction of the IC allows it to be placed in a small area. The IC regulator circuit can be used for low current and voltage levels.

Equipment

Electronic multifunction meter.
Transformer—6.3/12.6 V ac secondary.
Integrated circuit—μA723C.
Resistors—47,000 Ω, 10,000 Ω, 8200 Ω, 4700 Ω, 1000 Ω, 500 Ω, 250 Ω.
Capacitors—500 μF, 25 V dcw; 25 μF, 25 V dcw; 100 pF.
Diodes (2)—1N4004 or equivalent.
Potentiometer—10,000 Ω.

Procedure

1. Construct the IC voltage-regulator circuit shown in Fig. 3-26A.
2. Apply power to the circuit and measure the output voltage across the load terminals as R_2 is varied through its range of resistance.

   Minimum output voltage = ______ volts dc.

   Maximum output voltage = ______ volts dc.
3. Adjust R_2 until 7 volts dc appear across the load terminals. Do not readjust R_2 during the remainder of the activity.
4. Connect a 10,000-ohm resistor across the load terminals.

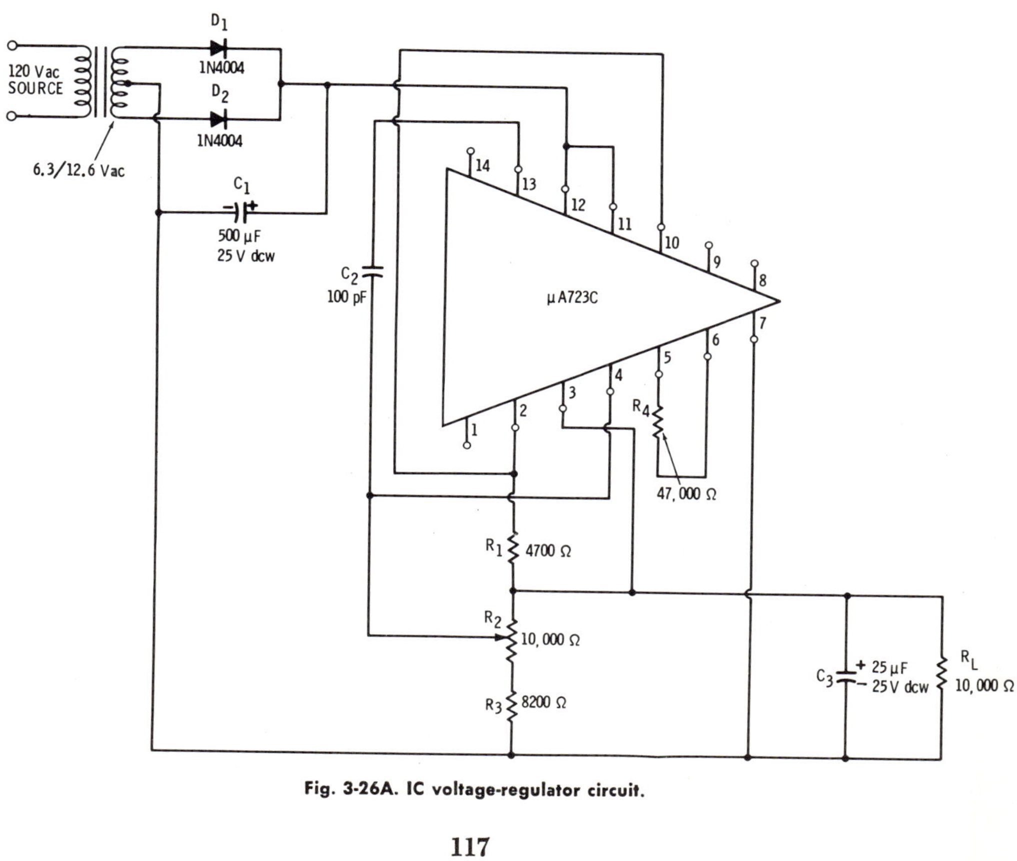

Fig. 3-26A. IC voltage-regulator circuit.

5. Complete the voltage and current measurements indicated in Table 3-26A as R_L is varied from 10,000 ohms to 180 ohms.
6. This concludes the experiment.

Analysis

1. What are some advantages of integrated-circuit voltage regulators over other types of regulators?

____

____

____

2. What is the purpose of capacitor C_3 in this circuit?

____

____

____

3. From the data of Table 3-26A, what is the voltage variation across C_1? Why?

____

____

____

Table 3-26A. IC Regulator Values

R_L	I_{R_L}	V_{C_1}	V_{R_L}
10,000 Ω			
1000 Ω			
500 Ω			
250 Ω			

4. As a greater amount of load is placed in the circuit (smaller resistance), how does I_{R_L} vary? How does V_{R_L} vary?

____

____

____

5. Calculate the voltage regulation of the circuit*:

$$\% \text{ Regulation} = \frac{V_{NL} - V_{FL}}{V_{FL}} \times 100 = \_\_\_\_\_\_ \%.$$

*Use V_{NL} as the 10,000-ohm load and V_{FL} as the 2500-ohm load.

4-1

Transformer Analysis

Objective

In this laboratory activity, you may use any multiple-secondary transformer and complete an analysis of its voltage and resistance characteristics. A multiple-secondary transformer has more than one secondary winding. You should note the voltage and current ratings of each winding. Be careful in making voltage measurements, since some secondary windings may be step-up windings while others may be step-down. Also, be careful not to short any of the secondary windings together since this may destroy the transformer.

Equipment

Multifunction meter.
Multiple-secondary transformer.
Variable ac source.

Procedure

1. The schematic diagram of a typical multiple-secondary transformer is shown in Fig. 4-1A. The transformer you use should be similar to this.

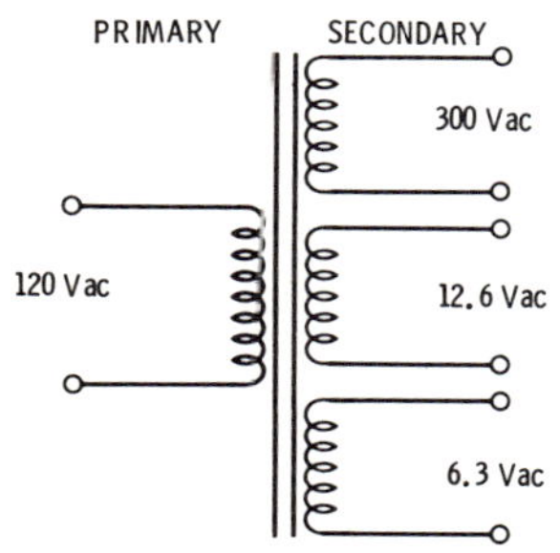

Fig. 4-1A. Schematic diagram of a typical multiple-secondary transformer.

2. Note the voltages and current ratings of the secondary windings, which may be obtained by applying the rated voltage to the primary winding. These are the rated values for this particular transformer. Note also whether the secondary windings are step-up or step-down. The transformer has _______ step-up windings and _______ step-down windings.

3. In the following space, draw a schematic diagram of the transformer you are using. Label the voltage and current rating for each winding if this information is available.

4. Apply the rated voltage to the primary winding and measure the voltage (V_s) across each secondary winding. Record these values in the blank spaces.

V_{s_1} = _______ volts ac.

V_{s_2} = _______ volts ac.

V_{s_3} = _______ volts ac.

V_{s_4} = _______ volts ac.

V_{s_5} = _______ volts ac.

5. Reduce the applied voltage to one-half its rated value and complete the same secondary measurements as in Step 4.

V_{s_1} = _______ volts ac.

V_{s_2} = _______ volts ac.

V_{s_3} = _______ volts ac.

V_{s_4} = _______ volts ac.

V_{s_5} = _______ volts ac.

6. With no power applied, make the following resistance measurements:

   Primary resistance (R_p) = ______ ohms.

   R_{s_1} = ______ ohms.

   R_{s_2} = ______ ohms.

   R_{s_3} = ______ ohms.

   R_{s_4} = ______ ohms.

   R_{s_5} = ______ ohms.

7. This concludes the experiment.

Analysis

1. What conclusion may be drawn from the resistance measurements of Step 6 concerning the current capacity of each secondary winding?

   ____

   ____

   ____

2. What is the relationship between primary power and secondary power in a multiple-secondary transformer?

   ____

   ____

   ____

3. Do the measured voltage values across the secondary windings correspond well with the rated values? Give an example to justify your answer.

   ____

   ____

   ____

4. Discuss the following characteristics of transformers:

   (a) Winding resistance.

   ____

   ____

   ____

   (b) Volt-ampere rating.

   ____

   ____

   ____

   (c) Core construction.

   ____

   ____

   ____

   (d) Magnetic coupling.

   ____

   ____

   ____

   (e) Turns ratio.

   ____

   ____

   ____

   (f) Voltage ratio.

   ____

   ____

   ____

5. Compare the voltage measurements of Steps 4 and 5.

   ____

   ____

   ____

4-2

Single-Phase Transformers—Impedance Transfer

Objective

Since the load is connected to the secondary winding of a transformer, the value of the load will affect the current drawn by the primary from the source. The primary current will provide the necessary induced secondary voltage to energize the load. This effect is referred to as *impedance transfer* from secondary to primary.

In this laboratory activity, you will be able to investigate impedance transfer in a transformer with different values of load resistance. You can use a commercially manufactured, dissectible transformer unit or a transformer that you have improvised in the lab. In the latter case, merely modify Table 4-2A so that the values correspond to those of the transformer you are using.

Equipment

Multifunction meter.
Variable ac voltage source.
Dissectible transformer or other appropriate transformer.
Ac current meter.
Resistors—10 Ω, 10 W; 100 Ω, 10 W; 500 Ω, 10 W.

Procedure

1. Construct the circuit shown in Fig. 4-2A.
2. Perform the necessary voltage and current measurements to complete the first three lines of Table 4-2A.

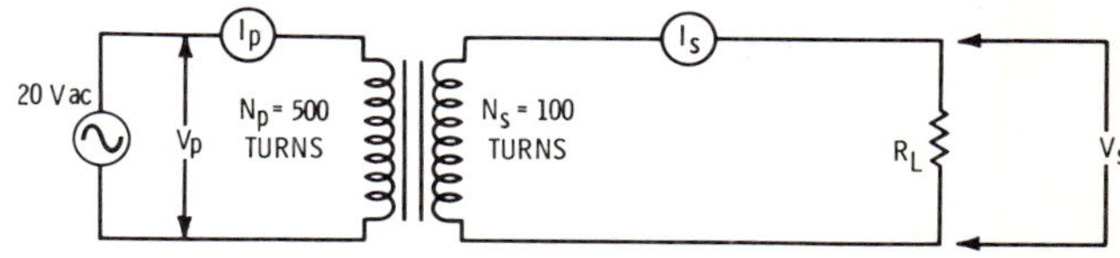

Fig. 4-2A. Circuit for observing the impedance transfer characteristics of a single-phase transformer.

   Begin with a load resistance of 10 ohms, then use a 100-ohm and a 500-ohm resistance.
3. Modify the secondary winding so that it now has 200 turns, and perform the measurements to complete lines 4–6 of Table 4-2A.
4. Modify the secondary winding so that it has 500 turns, and perform the measurements to complete lines 7–9 of Table 4-2A.
5. This concludes the experiment.

Analysis

1. Verify the following impedance transfer relationship from the data of lines 2, 4, and 9 of Table 4-2A:

$$Z_p = Z_s \times \left(\frac{N_p}{N_s}\right)^2 = \_\_\_\_\_\_\_\_ \text{ ohms.}$$

2. How does an increase in R_L affect:
   (a) I_p?

   ____

   (b) I_s?

   ____

Table 4-2A. Impedance Transfer Characteristics

R_L	N_s	I_p	$Z_p = V_p/I_p$*	V_s	I_s	$Z_s = V_s/I_s$	$(N_p/N_s)^2$
10 Ω	100						
100 Ω	100						
500 Ω	100						
10 Ω	200						
100 Ω	200						
500 Ω	200						
10 Ω	500						
100 Ω	500						
500 Ω	500						

* $V_p = 20$ Vac.

(c) Z_p?

(d) Z_s?

3. What is the relationship of Z_s and Z_p?

4. From the data of Table 4-2A, how does increasing the number of secondary turns affect:

(a) I_p?

(b) I_s?

(c) Z_p?

(d) Z_s?

Transformer Impedance Matching

Objective

A unique function that transformers can perform is known as *impedance matching*. We know that maximum power is transferred from source to load when the source impedance is equal to the load impedance. We can use transformers to match a given load impedance to a source impedance of a different value. This can be done by choosing a transformer with the proper turns ratio, since $Z_p = (N_p/N_s)^2 \times Z_s$. There are numerous applications of transformers as impedance matching devices in electronic equipment.

In this laboratory activity, you will observe the transfer of power from the source to the load by a transformer. You may use either a dissectible transformer unit or a commercially manufactured transformer.

Equipment

Multifunction meter.
Variable ac power source.
Dissectible transformer or other appropriate transformer.
Ac current meter.
Resistor—27 Ω, 10 W.
Potentiometer—1000 Ω.

Procedure

1. The connection diagram for a circuit to illustrate the impedance matching principle is shown in Fig. 4-3A. A 100-turn to 500-turn winding ratio is shown here; however, any ratio could be used.
2. Apply 10 volts ac to the primary of the transformer. Adjust the potentiometer to 50 ohms. Measure and record the values of V_{R_L}, I_p, and I_s. Record these values in Table 4-3A. Calculate and record the values of primary power (P_p), primary impedance (Z_p), secondary power (P_s), secondary impedance (Z_s), and turns ratio $(N_p/N_s)^2$.
3. Adjust the potentiometer through the range of values indicated in Table 4-3A, and complete each of the corresponding measurements and calculations.
4. This concludes the experiment.

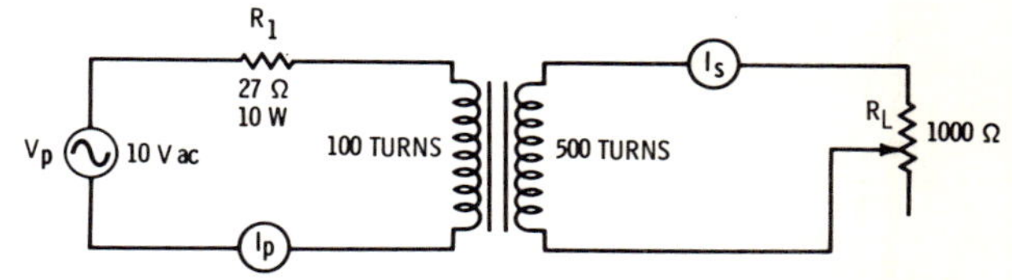

Fig. 4-3A. Circuit for observing the impedance matching principle of a transformer.

Analysis

1. What is meant by the term *impedance matching*?

2. How does a transformer accomplish impedance matching?

3. What is meant by maximum power transfer?

4. Prepare a graph as shown in the following space to illustrate R_L values versus secondary power (P_s).

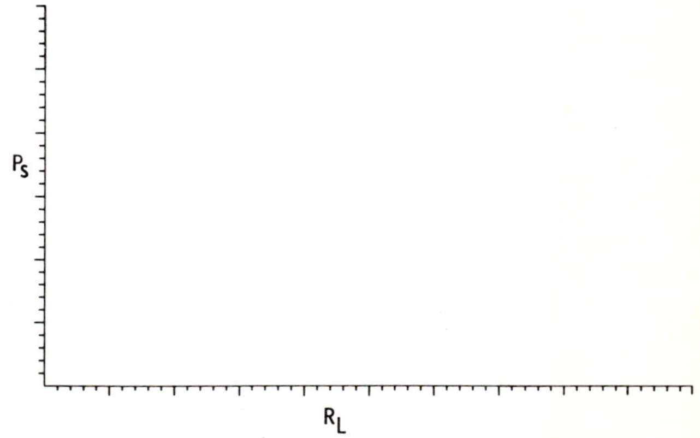

Table 4-3A. Impedance Matching Values

R_L	I_p	$P_p = V_p \times I_p$	$Z_p = V_p/I_p$	I_s	V_s	$P_s = V_s \times I_s$	$(N_p/N_s)^2$
50 Ω							
100 Ω							
200 Ω							
300 Ω							
400 Ω							
500 Ω							
600 Ω							
700 Ω							
800 Ω							
900 Ω							
1000 Ω							

5. At what value of R_L on the graph is maximum secondary power developed?

____

6. What is the relationship of maximum secondary power to primary power at the point of maximum power transfer?

____

____

____

7. What is a specific application where a transformer could be used as an impedance matching device?

____

Current Transformer Principle

Objective

Current transformers are often used to reduce a large value of line current to a smaller value for measurement or control purposes. These transformers are used to measure the current magnitude of high-current systems. Since most metering systems respond linearly to current changes, the current transformer principle could also be used to measure quantities other than current in high-power systems.

In this laboratory activity, you will observe the current transformer principle. You will be able to use any step-up transformer to illustrate this principle. You should note the difference in the values of primary current and secondary current throughout the experimental procedure.

Equipment

Multifunction meter.
Dissectible transformer or any step-up transformer.
Variable ac power source.
Ac current meter.
Potentiometer—1000 Ω.

Procedure

1. The connection diagram for observing the current transformer principle is shown in Fig. 4-4A. A 100-turn to 1000-turn ratio is used here; however, any step-up transformer could be used.
2. With 10 volts ac applied, vary the 1000-ohm potentiometer, beginning with 100 ohms, up to 1000 ohms. Record the necessary current values in Table 4-4A.
3. Apply 20 volts ac to the primary. Measure and record the primary and secondary current values as the potentiometer is varied.

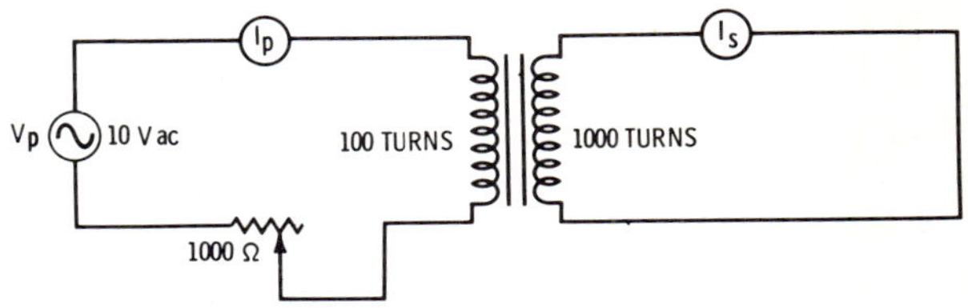

Fig. 4-4A. Circuit for observing the current transformer principle.

4. Perform the same procedure with 30 volts ac applied. Also record these values in Table 4-4A.
5. This concludes the experiment.

Analysis

1. Draw a schematic diagram of a current transformer performing a circuit measurement.

Table 4-4A. Primary and Secondary Current Comparison

Potentiometer Resistance	10 V ac		20 V ac		30 V ac	
	I_p	I_s	I_p	I_s	I_p	I_s
100 Ω						
200 Ω						
400 Ω						
600 Ω						
800 Ω						
1000 Ω						

2. What are some specific applications of current transformers?

______

______

______

3. From the data of Table 4-4A, discuss the variation of secondary current as primary current varies.

______

______

______

4. In reference to No. 3 of the analysis, how does the relationship of I_s and I_p illustrate the principle of current transformation?

______

______

______

5. From the data of Table 4-4A, what effect does increased voltage have on the relationship of I_s and I_p?

______

______

______

Autotransformer Principles

Objective

The autotransformer has only one winding; therefore, it requires less copper wire to construct than a conventional single-phase transformer of similar size. The major disadvantage of an autotransformer is the necessity of having a direct connection to one of the primary input lines.

In this laboratory activity, you will observe the operational principles of an autotransformer. You can use either an autotransformer or a secondary winding of a center-tapped transformer. You should notice that this device may either be constructed as a step-up or step-down transformer.

Equipment

Multifunction meter.
Variable ac power source.
Center-tapped transformer.
Resistors—10,000 Ω, 10 W; 1000 Ω, 10 W.
Ac current meter.

Procedure

1. Construct the autotransformer circuit shown in Fig. 4-5A.

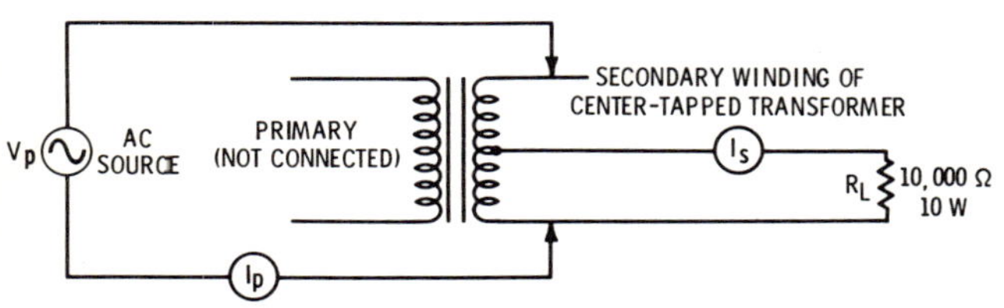

Fig. 4-5A. Autotransformer step-down connections.

2. Apply rated voltage to the outer winding terminals of the center-tapped transformer. Measure the voltage across the load (V_{R_L}), primary current (I_p), and secondary current (I_s).

   V_{R_L} = _______ volts ac.

   I_p = _______ amperes ac.

   I_s = _______ amperes ac.

3. Change the value of load resistance to 1000 ohms and repeat the voltage and current measurements.

   V_{R_L} = _______ volts ac.

   I_p = _______ amperes ac.

   I_s = _______ amperes ac.

4. Modify the circuit as shown in Fig. 4-5B. The ac input voltage should remain the same as before.

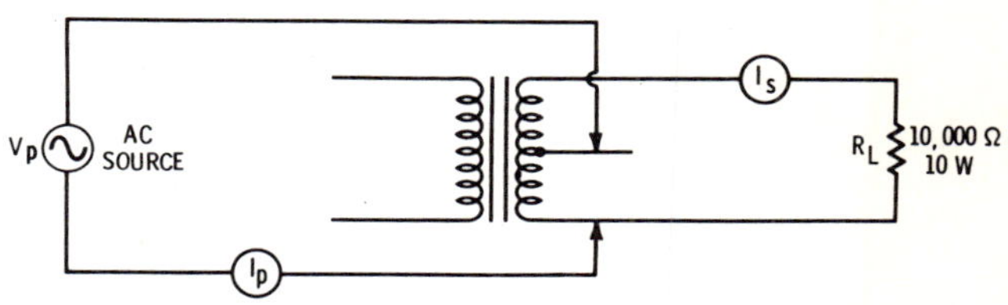

Fig. 4-5B. Autotransformer step-up connections.

5. Again, measure V_{R_L}, I_p, and I_s.

   V_{R_L} = _______ volts ac.

   I_p = _______ amperes ac.

   I_s = _______ amperes ac.

6. Change the value of load resistance to 1000 ohms and repeat the measurements.

   V_{R_L} = _______ volts ac.

   I_p = _______ amperes ac.

   I_s = _______ amperes ac.

7. This concludes the experiment.

Analysis

1. What type of autotransformer was constructed in Step 1?

   ____

2. What type of autotransformer was constructed in Step 4?

   ____

3. From Step 2, verify that $V_p \times I_p = V_s \times I_s$.

4. From Step 3, verify that $V_p \times I_p = V_s \times I_s$.

5. From Step 5, verify that $V_p \times I_p = V_s \times I_s$.

6. From Step 6, verify that $V_p \times I_p = V_s \times I_s$.

7. What are some applications of autotransformers?

8. How does decreased load resistance affect the primary current?

9. Are the primary and secondary power values ($V_p \times I_p$ and $V_s \times I_s$) equal? Why?

4-6

Three-Phase Transformer Connections

Objective

Three-phase transformers may be connected by using several different primary and secondary configurations. If the primary winding of a three-phase transformer is connected in a delta configuration and the secondary is wye-connected, the transformer is said to be connected in a delta-wye arrangement. It is also possible to connect three-phase transformers in a wye-wye, delta-delta, wye-delta, or open-delta configuration.

In this laboratory activity, you will be able to observe the characteristics of the five methods used to connect three-phase transformers. It is also possible to use three identical single-phase transformers if a three-phase transformer is not available.

Equipment

Three-phase transformer or three identical single-phase transformers.
Three-phase ac power source.
Multifunction meter.
Lamps (3)—60 W.

Procedure

1. Construct the wye-wye system shown in Fig. 4-6A. Be sure to observe polarity of the transformer windings. Use a wye-connected load consisting of three 60-watt lamps.
2. Apply ______ volts from the three-phase ac power source.
3. Make the necessary voltage and current measurements to complete Table 4-6A.
4. Turn off the power source and reconnect the windings in a delta-delta arrangement as shown in Fig. 4-6B. Keep the voltage source at the same value on this and each of the other connections.
5. Make the necessary voltage and current measurements to complete Table 4-6B.
6. Turn off the power source. Refer to the two previous connection diagrams and reconnect the transformer windings in a wye-delta configuration with a delta-connected load. It will only be necessary to modify the primary winding from the previous connection method.
7. Make the necessary measurements to complete Table 4-6C.
8. Turn off the power source. Reconnect the transformer windings in a delta-wye configuration with a wye-connected load. Refer to the previous diagrams.
9. Make the necessary measurements to complete Table 4-6D.
10. Turn off the power source. Reconnect the transformer secondary windings in an open-delta configuration. The primary will remain connected in delta. The load should be delta-connected as shown in Fig. 4-6C.
11. Make the necessary measurements to complete Table 4-6E.
12. This concludes the experiment.

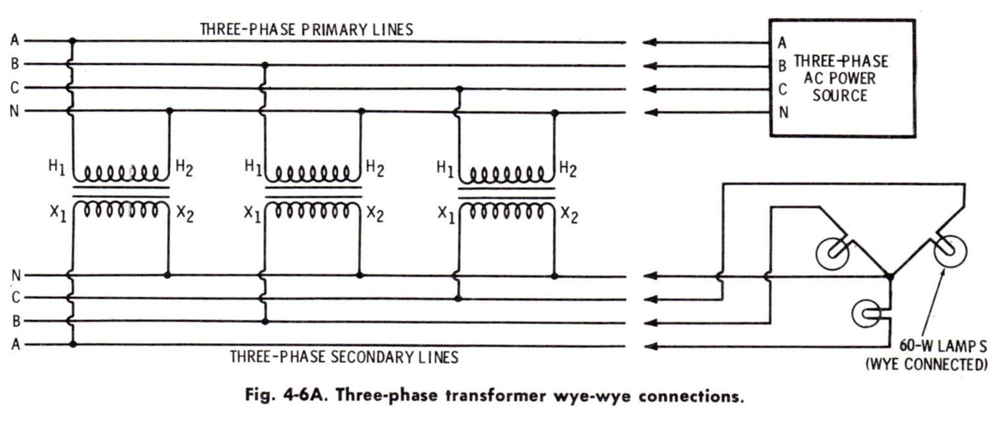

Fig. 4-6A. Three-phase transformer wye-wye connections.

Table 4-6A. Wye-Wye Connection

Primary Measurements								
$V_{A\text{-}B}$	$V_{B\text{-}C}$	$V_{A\text{-}C}$	$V_{A\text{-}N}$	$V_{B\text{-}N}$	$V_{C\text{-}N}$	I_A	I_B	I_C

Secondary Measurements								
$V_{A\text{-}B}$	$V_{B\text{-}C}$	$V_{A\text{-}C}$	$V_{A\text{-}N}$	$V_{B\text{-}N}$	$V_{C\text{-}N}$	I_A	I_B	I_C

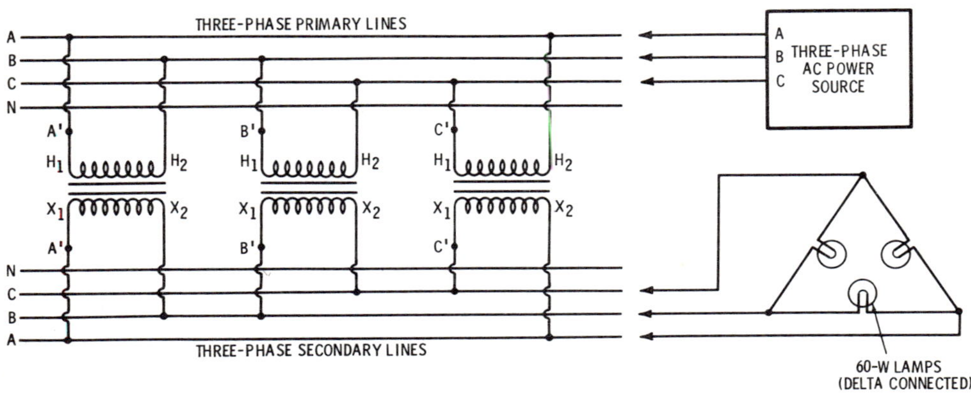

Fig. 4-6B. Three-phase transformer delta-delta connections.

Table 4-6B. Delta-Delta Connection

Primary Measurements								
$V_{A\text{-}B}$	$V_{B\text{-}C}$	$V_{A\text{-}C}$	$I_{L(A)}$*	$I_{L(B)}$*	$I_{L(C)}$*	$I_{p(A)}$†	$I_{p(B)}$†	$I_{p(C)}$†

Secondary Measurements								
$V_{A\text{-}B}$	$V_{B\text{-}C}$	$V_{A\text{-}C}$	$I_{L(A)}$*	$I_{L(B)}$*	$I_{L(C)}$*	$I_{p(A)}$†	$I_{p(B)}$†	$I_{p(C)}$†

* Line current in phases A, B, and C.
† Phase current in phases A, B, and C (points A′, B′, and C′ on the diagram).

Table 4-6C. Wye-Delta Connection

Primary Measurements								
$V_{A\text{-}B}$	$V_{B\text{-}C}$	$V_{A\text{-}C}$	$V_{A\text{-}N}$	$V_{B\text{-}N}$	$V_{C\text{-}N}$	I_A	I_B	I_C

Secondary Measurements								
$V_{A\text{-}B}$	$V_{B\text{-}C}$	$V_{A\text{-}C}$	$I_{L(A)}$*	$I_{L(B)}$*	$I_{L(C)}$*	$I_{p(A)}$†	$I_{p(B)}$†	$I_{p(C)}$†

* Line current in phases A, B, and C.
† Phase current in phases A, B, and C.

Table 4-6D. Delta-Wye Connection

Primary Measurements								
$V_{A\text{-}B}$	$V_{B\text{-}C}$	$V_{A\text{-}C}$	$I_{L(A)}$*	$I_{L(B)}$*	$I_{L(C)}$*	$I_{p(A)}$†	$I_{p(B)}$†	$I_{p(C)}$†
Secondary Measurements								
$V_{A\text{-}B}$	$V_{B\text{-}C}$	$V_{A\text{-}C}$	$V_{A\text{-}N}$	$V_{B\text{-}N}$	$V_{C\text{-}N}$	I_A	I_B	I_C

* Line current in phases A, B, and C.
† Phase current in phases A, B, and C.

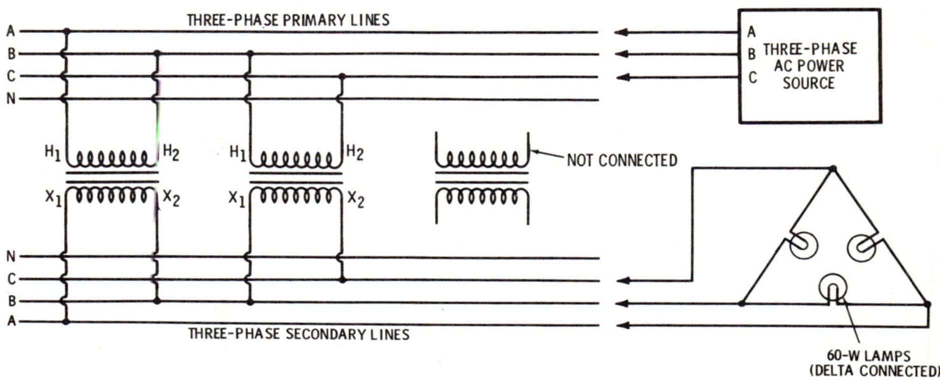

Fig. 4-6C. Three-phase transformer open-delta connections.

Table 4-6E. Open-Delta Connection

Primary Measurements								
$V_{A\text{-}B}$	$V_{B\text{-}C}$	$V_{A\text{-}C}$	$I_{L(A)}$*	$I_{L(B)}$*	$I_{L(C)}$*	$I_{p(A)}$†	$I_{p(B)}$†	$I_{p(C)}$†
Secondary Measurements								
$V_{A\text{-}B}$	$V_{B\text{-}C}$	$V_{A\text{-}C}$	$I_{L(A)}$*	$I_{L(B)}$*	$I_{L(C)}$*	$I_{p(A)}$†	$I_{p(B)}$†	$I_{p(C)}$†

* Line current in phases A, B, and C
† Phase current in phases A, B, and C.

Analysis

1. What would happen if the polarity of one of the primary transformer windings was not observed?

____

____

2. From the data of Table 4-6A, compare:
   (a) $V_{A\text{-}B}$ to $V_{A\text{-}N}$ of the primary.

   ____

   ____

   (b) $V_{A\text{-}B}$ of the primary to $V_{A\text{-}B}$ of the secondary.

   ____

   ____

   (c) I_A in the primary to I_A in the secondary.

   ____

   ____

   (d) $V_{A\text{-}B}$ to $V_{A\text{-}N}$ of the secondary.

   ____

   ____

3. From the data of Table 4-6B, compare:
   (a) V_{A-B} of the primary to V_{A-B} of the secondary.
   (b) $I_{L(A)}$ to $I_{p(A)}$ of the primary.
   (c) $I_{L(A)}$ to $I_{p(A)}$ of the secondary.
   (d) $I_{L(A)}$ of the primary to $I_{L(A)}$ of the secondary.
   (e) $I_{p(A)}$ of the primary to $I_{p(A)}$ of the secondary.

4. From the data of Table 4-6C, compare:
   (a) V_{A-B} to V_{A-N} of the primary.
   (b) I_A of the primary to $I_{L(A)}$ of the secondary.
   (c) I_A of the primary to $I_{p(A)}$ of the secondary.
   (d) V_{A-B} of the secondary to V_{A-B} of the secondary from Table 4-6B.
   (e) V_{A-B} of the secondary to V_{A-B} of the secondary from Table 4-6A.

5. From the data of Table 4-6D, compare:
   (a) V_{A-B} of the primary to V_{A-B} of the secondary.
   (b) V_{A-B} to V_{A-N} of the secondary.
   (c) $I_{L(A)}$ to $I_{p(A)}$ of the primary.
   (d) $I_{L(A)}$ of the primary to I_A of the secondary.
   (e) $I_{p(A)}$ of the primary to I_A of the secondary.
   (f) V_{A-B} of the secondary to V_{A-B} of the secondary from Table 4-6C.

6. From the data of Table 4-6E, compare:
   (a) V_{A-B} of the primary to V_{A-B} of the secondary.
   (b) $I_{L(A)}$ of the primary to $I_{L(A)}$ of the secondary.
   (c) $I_{p(A)}$ of the primary to $I_{p(A)}$ of the secondary.
   (d) V_{A-B} of the secondary to V_{A-B} of the secondary from Table 4-6B.
   (e) $I_{L(A)}$ of the secondary to $I_{L(A)}$ of the secondary from Table 4-6B.
   (f) $I_{p(A)}$ of the secondary to $I_{p(A)}$ of the secondary from Table 4-6B.

Electrical Conductors and Insulators

Objective

Two very important considerations in industrial distribution systems are conductors and insulators. Conductors of various types are designed to carry electrical current. Conductors may be in the form of solid wires, stranded cables, cords, bus bars, or many other types. The amount of current that a conductor can safely handle is mainly determined by its diameter. Two examples of conductor tables are included in Appendix 1.

The diameter of a conductor is measured with an American Wire Gauge (AWG). The sizes from No. 14 through 1000 MCM (1,000,000 circular mils) are shown in Table A1-2. Conductors are made of either copper or aluminum.

The insulation used on electrical conductors is mainly determined by the type of location where it will be used and the type of enclosure or wireway where it will be placed. The symbols R, H, W, T, and N indicate rubber, heat-resistant, moisture-resistant, thermoplastic and nylon, respectively. A table of insulation types is shown in Appendix 2.

In this activity, you will observe several types of conductors and insulation. You should use the samples of conductors provided by your instructor.

Equipment

American Wire Gauge (AWG).
Conductor samples.

Procedure

1. Obtain an American Wire Gauge and several conductors from your instructor.
2. Measure the diameter of each conductor with the AWG and record in the following spaces. Also record whether each conductor is copper or aluminum.
   Sizes of conductors:

   (1) ______________.

   (2) ______________.

   (3) ______________.

   (4) ______________.

   (5) ______________.

   (6) ______________.
3. Refer to the tables in Appendix 1 and determine the amount of current each conductor can carry. Record in the following spaces:
   Ampacity of conductors:

   (1) _______ amperes.

   (2) _______ amperes.

   (3) _______ amperes.

   (4) _______ amperes.

   (5) _______ amperes.

   (6) _______ amperes.
4. Now determine the type of insulation on each and record in the following spaces:
   Type of insulation:

   (1) ______________.

   (2) ______________.

   (3) ______________.

   (4) ______________.

   (5) ______________.

   (6) ______________.

Analysis

1. What is the area in circular mils of the following conductors?

   (a) No. 16 = _______ cmil.

   (b) No. 0 = _______ cmil.

   (c) No. 8 = _______ cmil.

   (d) 250 MCM = _______ cmil.

   (e) 1000 MCM = _______ cmil.

   (f) No. 12 = _______ cmil.
2. How many wires are used to construct the conductors listed in the preceding?

   (a) _______.

(b) ______.

(c) ______.

(d) ______.

(e) ______.

(f) ______.

3. What are the types of the following insulations?

(a) RH = ____________.

(b) RHW = ____________.

(c) THW = ____________.

4. What are the ampacities of the following conductors?

(a) No. 12 (RH) copper = ______ amperes.

(b) No. 4 (TW) aluminum = ______ amperes.

(c) 250 MCM (THW) aluminum = ______ amperes.

(d) No. 6 (T) copper = ______ amperes.

(e) No. 000 (R) aluminum = ______ amperes.

(f) 1000 MCM copper = ______ amperes.

5. What is the dc resistance of the following lengths of conductors?

(a) 500 feet of No. 8 copper = ______ ohms.

(b) 800 feet of No. 12 aluminum = ______ ohms.

(c) 5000 feet of 250 MCM aluminum = ______ ohms.

(d) 1500 feet of 500 MCM copper = ______ ohms.

(e) 400 feet of No. 6 copper = ______ ohms.

(f) 8000 feet of 1000 MCM aluminum = ______ ohms.

4-8

Printed Circuit Construction

Objective

When working with industrial electrical systems, it is often desirable to design and construct circuits for personal use or use in the laboratory. Many types of circuits can be constructed using printed circuit (pc) boards. This activity outlines some methods that can be used to make printed circuit boards.

There are some basic fundamentals of pc boards of which the technician should be aware. Actual circuit construction is limited only by the ingenuity of the designer. The following statements are beneficial to pc board design:

1. Draw a simple schematic of the circuit you wish to construct.
2. Obtain the necessary components to build the circuit, and all other equipment and materials (copper-clad board, soldering pencil, solder, needle-nose pliers, cutting pliers, etc.) you will need.
3. Keep the width of the copper conductive paths about 1/16-inch wide and the rounded terminal ends at least 1/16-inch apart.

In this laboratory activity, the procedures for making printed circuits will be presented. This activity could either be used for individuals or groups for the construction of printed circuits.

Equipment

Copper-clad board (size according to circuit design).
Ferric-chloride etching solution.
Glass tray.
Steel wool.
Carbon paper and pencil.
Acid-resist pen, tape, or paint.
Soap and water.
Wooden stick.

Procedure

Section A. Acid-Resist Pen Process

1. Clean the copper side of the copper-clad board with steel wool. Grasp the board by the edges to avoid touching the copper when it is clean.
2. Use soap and water to wash the board, then dry it with a cloth or paper towel.
3. Place the circuit layout diagram you have designed on the copper side of the copper-clad board. Place a piece of carbon paper between the layout diagram and the copper-clad board.
4. Trace the circuit diagram completely with a pencil to transfer it to the copper-clad board. Remove the carbon paper and layout diagram.
5. Carefully and accurately trace over the diagram on the copper-clad board with the acid-resist pen.
6. After allowing time for drying, place the prepared board in a tray.
7. Pour enough ferric-chloride etching solution into the tray to cover the board. Be *very* careful with this chemical solution.
8. The board must be moved to avoid copper crystallization on the board. You can use a wooden stick to move the board. The ferric-chloride solution will etch several boards; however, its etching action slows with age. Occasionally, it may be necessary to add more solution to speed the process.
9. Remove the board from the solution when all excess copper has been removed. Do not touch the solution.
10. Clean the board with soap and water and dry it with a cloth or paper towel.
11. Use lacquer thinner to remove the resist coating from the board, then clean the copper strips with steel wool.
12. Drill holes into the ends of the copper strips for component mounting.
13. Mount and solder the components used in your circuit design.

Section B. Tape-Resist Process

1. Complete Steps 1–4 of Section A to prepare the copper-clad board.
2. Carefully and accurately cut out pieces of electrical tape to the correct width of the conducting strips. Press these strips firmly onto the outlined areas of the board.
3. Cut out circular end pieces of tape and press them onto the ends of the conductive strips.

4. Complete Steps 6–13 of Section A to finish the printed circuit construction.

*Section C. Photoresist Process**

1. Make a photographic negative of the circuit layout diagram. The diagram should be on a clear plastic sheet or tracing paper.
2. Place the negative on the commercially prepared photoresist board.
3. Expose the board to light through the negative for about 10 minutes.
4. Place the board in a tray and pour enough developing solution to cover it. Leave the board in the solution for about 5 minutes, or until the circuit lines are distinct.
5. Remove the board and rinse it with water. Do not touch the copper side of the board.
6. Complete Steps 6–13 of Section A to finish the printed circuit construction.

Section D. Paint-Resist Process

1. Complete Steps 1–4 of Section A to prepare the copper-clad board.
2. Adequately mix a can of acid-resist paint.
3. Use a small paint brush to apply the resist paint to the circuit conductive strips. Be careful and accurate to produce an attractive circuit board.
4. Place the painted board near a hot plate or heat lamp to dry the paint.
5. After the paint is dry, complete Steps 6–13 of Section A to finish the printed circuit construction.

*Requires special photoresist board, developing solution, heat source, and photographic equipment.

Analysis

1. What are some advantages of printed circuits?

____

____

____

2. Which method of pc board construction would be most desirable for mass production? Why?

____

____

____

3. Discuss briefly each of the four printed circuit construction methods that were outlined in this activity.

____

____

____

____

____

____

____

____

____

____

____

____

4-9

Soldering and Terminal Connection Exercises

Objective

In this activity, you will demonstrate your skill in three important exercises: (1) soldering, (2) desoldering, and (3) terminal connection. An understanding of the proper techniques for performing these exercises is needed by technicians who design, service, or repair industrial equipment.

Equipment

Soldering iron—25 to 50 W.
Desoldering tool.
Solder.
No. 22 insulated wire—3-inch length.
Printed circuit board.
Component mounting strip.
Resistor.
Wire stripper.
Side-cutting pliers.
Needle-nose pliers.
"Solderless" connector and wire.
Crimping tool.

Procedure

Section A. Soldering Iron Preparation and Component Mounting

1. Allow the soldering iron to heat to its operating temperature.
2. Apply a small amount of solder to tin the tip of the iron.
3. Obtain a component-mounting strip and a resistor from your instructor.
4. Using proper soldering techniques, solder the resistor onto the component mounting strip.

Instructor's Approval: ______________________

Section B. Printed Circuit Board Soldering

1. Obtain a printed circuit (pc) board and a 3-inch length of No. 22 insulated wire.
2. Place the wire near the holes in the pc board where you intend to solder it to determine the length of wire required.
3. Strip ⅜ inch to ½ inch of insulation from one end of the wire with the wire stripper.
4. Cut off any excess wire and strip the insulation off the other end of the wire. The wire should lie flat on the pc board when soldered.
5. Using the proper soldering techniques, solder both ends of the wire to the pc board terminals.

Instructor's Approval: ______________________

Section C. Printed Circuit Board Desoldering

1. Obtain a desoldering tool.
2. Allow the desoldering tool to heat to its operating temperature.
3. Using the proper desoldering techniques, remove the piece of wire from the pc board without damaging the conductive strips on the pc board.

Instructor's Approval: ______________________

Section D. Installation of Terminal Connection

1. Obtain a "solderless" terminal connector crimping tool, and a length of insulated wire from your instructor.
2. Strip enough insulation from the wire so that the terminal connector fits properly.
3. Use the crimping tool to fasten the terminal connector to the wire.

Instructor's Approval: ______________________

5-1

Series Resistive-Inductive (RL) Circuits

Objective

Series RL circuits are encountered in many industrial systems. When an ac voltage is applied to this type of circuit, the current will be the same through each component. However, the voltage drops across each component will be distributed according to the relative value of resistance and inductive reactance in the circuit. The vector sum of the voltage drops will equal the source voltage. Therefore, in calculating voltage distribution in the circuit, we must use the following formula:

$$V_T = \sqrt{V_R^2 + V_L^2}$$

where,

V_T is the total applied voltage,
V_R is the voltage across the resistance,
V_L is the voltage across the inductance.

In this laboratory activity, you will observe the characteristics of a series ac circuit that has resistance and inductance. The inductance characteristic is associated with many industrial systems that have windings, such as motors, generators, and transformers.

Equipment

Ac voltage source.
Resistor—10,000 Ω.
Inductor—8 H.
Electronic multifunction meter.

Procedure

1. Construct the series RL circuit shown in Fig. 5-1A.
2. Calculate the following:
   (a) Inductive reactance (X_L) = $2\pi fL$ =

   ________ ohms.

   (b) Impedance (Z) = $\sqrt{R^2 + X_L^2}$ =

   ________ ohms.
3. With a meter, measure the following values:
   (a) Voltage across the resistor (V_R) =

   ________ volts ac.

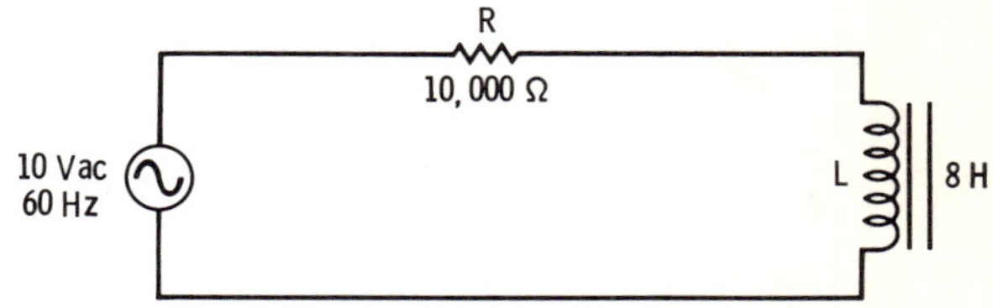

Fig. 5-1A. Series RL circuit.

   (b) Voltage across the inductor (V_L) =

   ________ volts ac.

   (c) Current through the circuit (I_T) =

   ________ amperes ac.
4. Experimentally determine the impedance of the circuit using the following method:

   (a) $I_T = \frac{V_R}{R}$ = ________ amperes.

   (b) $Z_T = \frac{V_T}{I_T}$ = ________ ohms.
5. Complete the following impedance triangle for the circuit you constructed by using the calculated values of R, X_L, and Z:

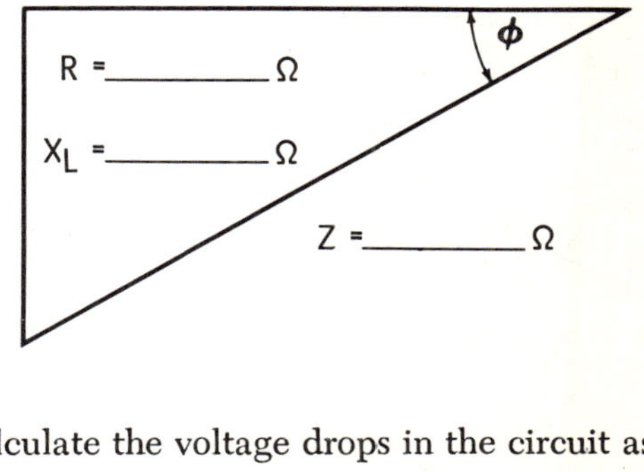

6. Calculate the voltage drops in the circuit as:

   (a) $V_R = I_T \times R$ = ________ volts ac.

   (b) $V_L = I_T \times X_L$ = ________ volts ac.
7. Show that the voltages must be added vectorially by using the values you calculated:

   $V_T = \sqrt{V_R^2 + V_L^2}$ = ________ volts ac.

8. Complete the following voltage triangles for this circuit using the calculated, then the measured values of V_T, V_R, and V_L:

CALCULATED VALUES

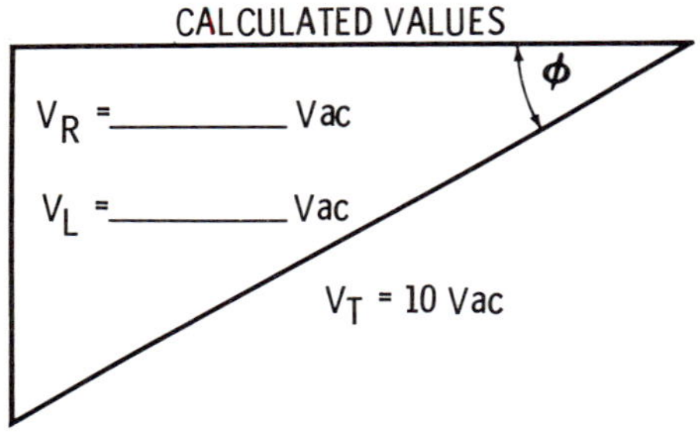

MEASURED VALUES

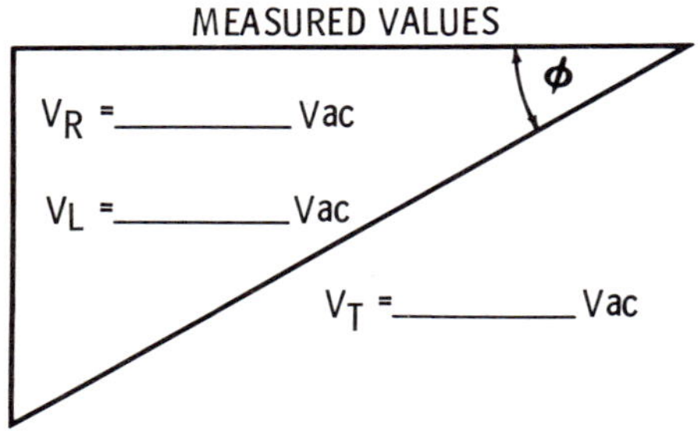

9. Calculate the power converted in the circuit:
   (a) True power (watts) $= I_T \times V_R =$

   _______ watts.

   (b) Apparent power (voltamperes) $= I_T \times V_T =$

   _______ voltamperes.

   (c) Reactive power (var) $= I_T \times V_L =$

   _______ var.

10. Complete the following power triangle for the circuit using your calculated values of true power, apparent power, and reactive power:

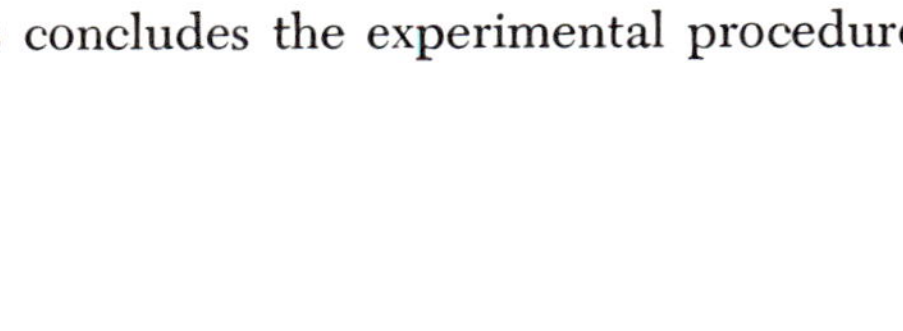

11. This concludes the experimental procedure.

Analysis

1. Why must the voltage in a series RL circuit be added vectorially?

____

____

____

2. Compare the calculated and measured values of the following:

Value	Calculated	Measured
I_T		
Z		
V_R		
V_L		

3. What are some factors that could cause a difference in your measured values and the calculated values?

____

____

____

4. Determine the phase angle of the circuit by using the following trigonometric methods:

   (a) Phase angle $(\phi) = \text{arc tangent } \frac{X_L}{R} =$ _______.

   (b) Phase angle $(\phi) = \text{arc tangent } \frac{V_L}{V_R} =$ _______.

   (c) Phase angle $(\phi) = \text{arc tangent } \frac{\text{var}}{\text{W}} =$ _______.

5. Determine the power factor of the circuit as:

$$\text{Power factor (pf)} = \frac{\text{true power}}{\text{apparent power}} = \_\_\_\_\_\_.$$

6. Determine the cosine of the phase angle (ϕ). This value should be equal to the circuit power factor (pf = cosine ϕ).

   Cosine $\phi =$ _____________.

Series Resistive-Capacitive (RC) Circuits

Objective

Series RC circuits are used in many industrial systems. When an ac voltage is applied to this type of circuit, the characteristics are similar to those of a series RL circuit. You should recall that in a capacitive circuit, current leads voltage (leading phase angle) and in an inductive circuit, voltage leads current (lagging phase angle). Therefore, the vectorial relationships show the reactive values in the opposite direction.

In this laboratory activity, you will observe the characteristics of a series ac circuit that has resistance and capacitance.

Equipment

Ac voltage source.
Resistor—10,000 Ω.
Capacitor—1.0 μF.
Electronic multifunction meter.

Procedure

1. Construct the series RC circuit shown in Fig. 5-2A.

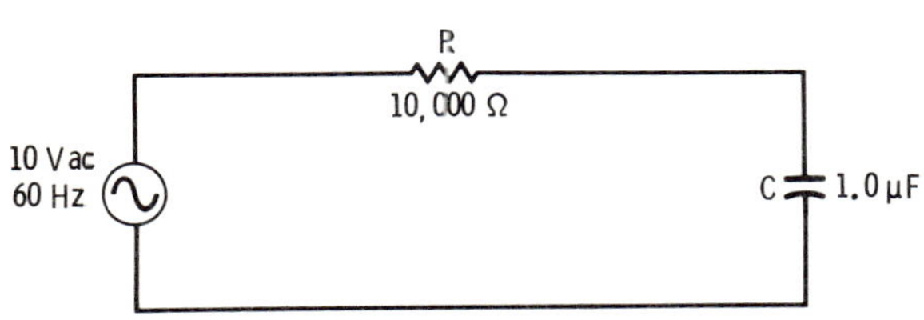

Fig. 5-2A. Series RC circuit.

2. Calculate the following:

   (a) Capacitive reactance $(X_C) = \frac{1}{2\pi fC} =$

   ______ ohms.

   (b) Impedance $(Z) = \sqrt{R^2 + X_C^2} =$

   ______ ohms.

3. With a meter, measure the following values:

   (a) Voltage across the resistor $(V_R) =$

   ______ volts ac.

   (b) Voltage across the capacitor $(V_C) =$

   ______ volts ac.

   (c) Current through the circuit $(I_T) =$

   ______ amperes ac.

4. Experimentally determine the impedance of the circuit using the following method:

   (a) $I_T = \frac{V_R}{R} =$ ______ amperes.

   (b) $Z_T = \frac{V_T}{I_T} =$ ______ ohms.

5. Complete the following impedance triangle for the circuit you constructed by using the calculated values of R, X_C, and Z:

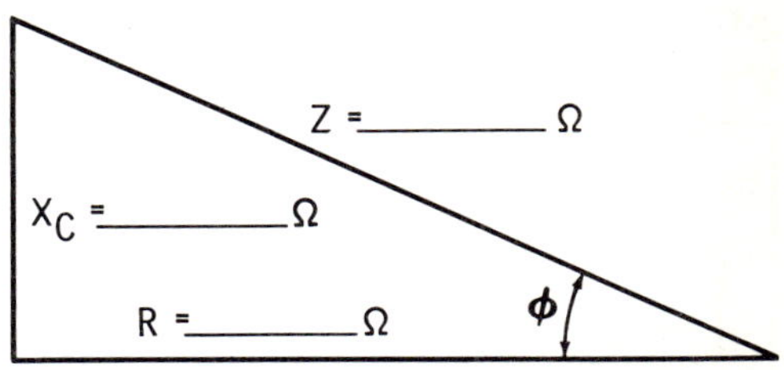

6. Calculate the voltage drops in the circuit as:

   (a) $V_R = I_T \times R =$ ______ volts ac.

   (b) $V_C = I_T \times X_C =$ ______ volts ac.

7. Show that the voltages must be added vectorially by using the values you calculated:

   $V_T = \sqrt{V_R^2 + V_C^2} =$ ______ volts ac.

8. Complete the following voltage triangles for this circuit using the calculated, then the measured values of V_T, V_R, and V_C:

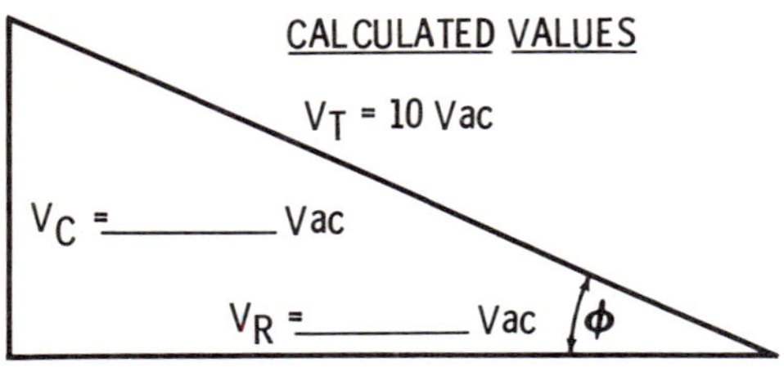

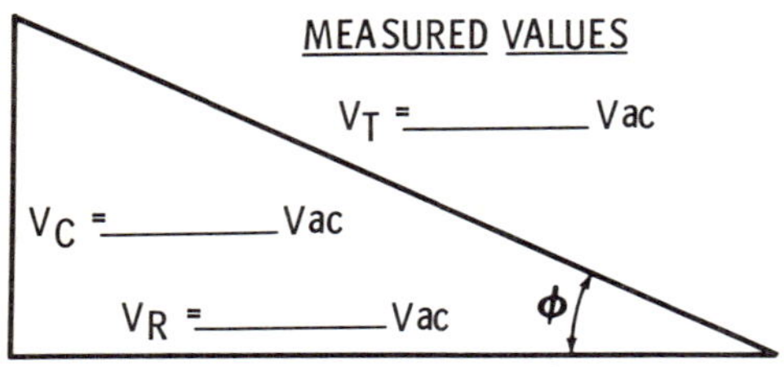

9. Calculate the power converted in the circuit:
   (a) True power (watts) $= I_T \times V_R =$

   ______ watts.
   (b) Apparent power (voltamperes) $= I_T \times V_T =$

   ______ voltamperes.
   (c) Reactive power (var) $= I_T \times V_C =$

   ______ var.
10. Complete the following power triangle for the circuit by using your calculated values of true power, apparent power, and reactive power:

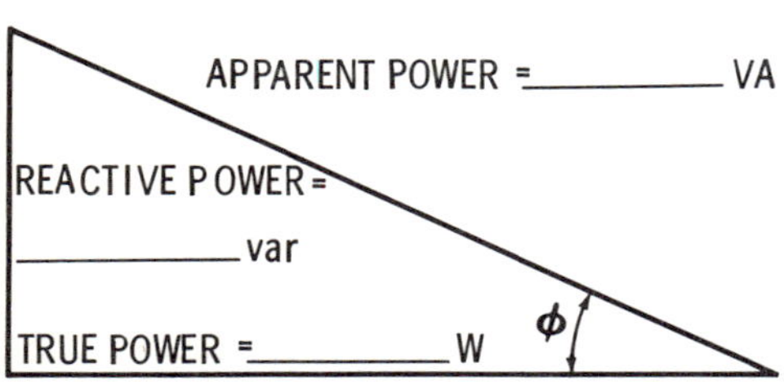

11. This concludes the experimental procedure.

Analysis

1. Compare the calculated and measured values of the following:

Value	Calculated	Measured
I_T		
Z		
V_R		
V_C		

2. Determine the phase angle of the circuit by using the following trigonometric methods:

   (a) Phase angle $(\phi) = \text{arc tangent}\,\frac{X_C}{R} =$ ______.

   (b) Phase angle $(\phi) = \text{arc tangent}\,\frac{V_C}{V_R} =$ ______.

   (c) Phase angle $(\phi) = \text{arc cosine}\,\frac{W}{VA} =$ ______.

3. Determine the power factor of the circuit as:

   $$\text{Power factor (pf)} = \frac{\text{true power}}{\text{apparent power}} = \text{\_\_\_\_\_\_}.$$

4. Determine the cosine of the phase angle. This value should equal the power factor of the circuit.

   Cosine $\phi =$ ______.

Series RCL Circuits

Objective

In this laboratory activity, you will observe the characteristics of an ac circuit that has resistive, capacitive, and inductive components. Since the effects of the capacitance and inductance are 180° out of phase with each other, we must analyze these circuits by using their vectorial relationships.

Equipment

Ac voltage source.
Resistor—10,000 Ω.
Capacitor—1.0 μF.
Inductor—8 H.
Electronic multifunction meter.

Procedure

1. Construct the series RCL circuit shown in Fig. 5-3A.

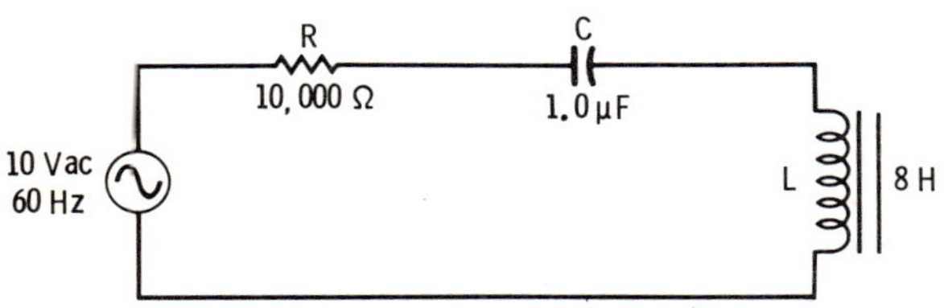

Fig. 5-3A. Series RCL circuit.

2. Calculate the following:
   (a) Capacitive reactance (X_C) $= \dfrac{1}{2\pi fC}$ = ________ ohms.
   (b) Inductive reactance (X_L) $= 2\pi fL =$ ________ ohms.
   (c) Total reactance (X_T) $= X_L - X_C =$ ________ ohms.
   (d) Impedance (Z) $= \sqrt{R^2 + (X_L - X_C)^2} =$ ________ ohms.
3. With a meter, measure the following values:
   (a) Voltage across the capacitor (V_C) = ________ volts ac.
   (b) Voltage across the inductor (V_L) = ________ volts ac.
   (c) Voltage across the resistor (V_R) = ________ volts ac.
   (d) Current through the circuit (I_T) = ________ amperes ac.
4. Experimentally determine the impedance of the circuit using the following method:
   (a) $I_T = \dfrac{V_R}{R} =$ ________ amperes.
   (b) $Z_T = \dfrac{V_T}{I_T} =$ ________ ohms.
5. Complete the following impedance triangle for the circuit you constructed by using the calculated values of R, X_C, X_L, and Z:

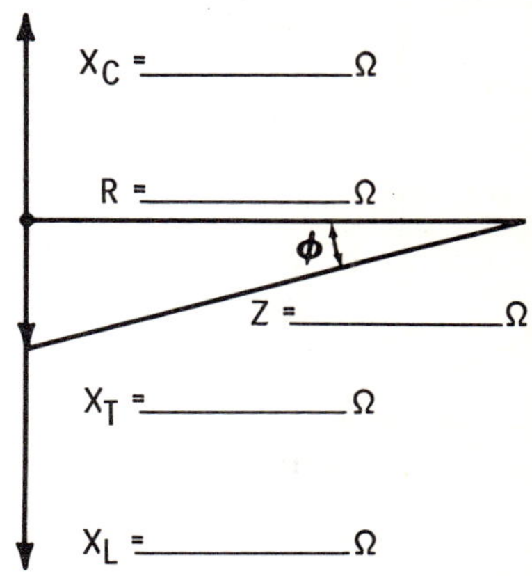

6. Calculate the voltage drops in the circuit as:
   (a) $V_R = I_T \times R$ = ________ volts ac.
   (b) $V_C = I_T \times X_C$ = ________ volts ac.
   (c) $V_L = I_T \times X_L$ = ________ volts ac.
   (d) $V_X = V_L - V_C$ = ________ volts ac.
7. Show that the voltages must be added vectorially by using the values you calculated:

$V_T = \sqrt{V_R^2 + (V_L - V_C)^2} =$ ________ volts ac.

8. Complete the following voltage triangles for this circuit using the calculated, then the measured values of V_T, V_R, V_L, and V_C.

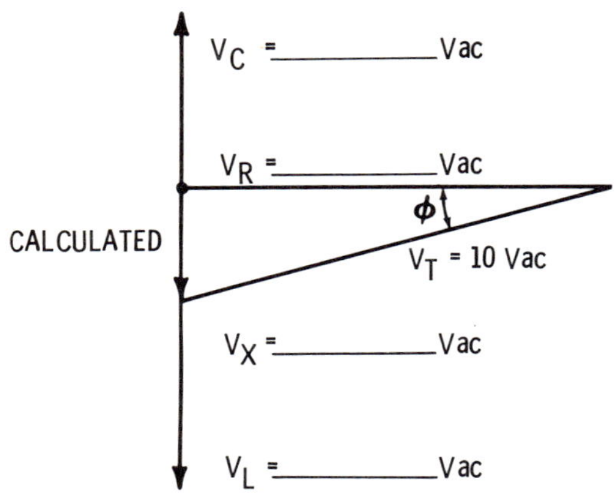

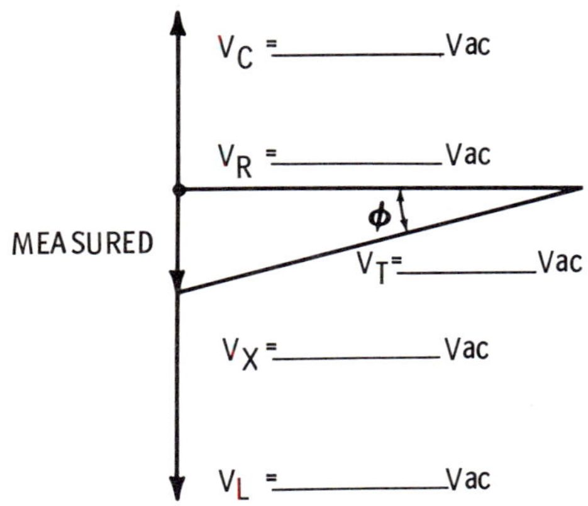

9. Calculate the power converted in the circuit:
   (a) True power (watts) $= I_T \times V_R =$

   ______ watts.
   (b) Apparent power (voltamperes) $= I_T \times V_T =$

   ______ voltamperes.
   (c) Capacitive reactive power (var_C) $= I_T \times V_C =$

   ______ var.
   (d) Inductive reactive power (var_L) $= I_T \times V_L =$

   ______ var.
   (e) Total reactive power (var_T) $= I_T \times V_X =$

   ______ var.

10. Complete the following power triangle for the circuit by using your calculated values of true power, apparent power, and reactive power:

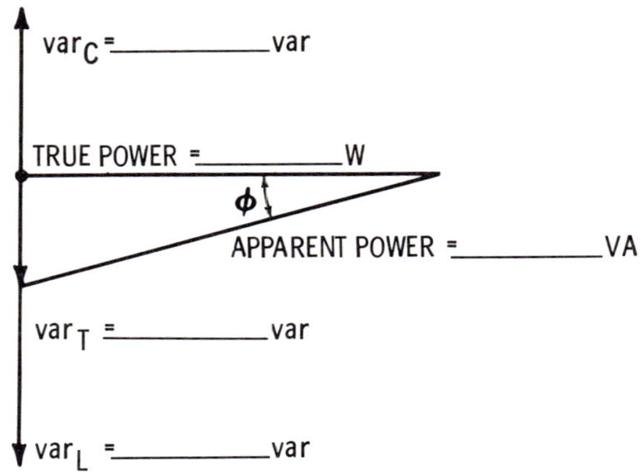

11. This concludes the experiment.

Analysis

1. Compare the calculated and measured values of the following:

Value	Calculated	Measured
I_T		
Z		
V_R		
V_C		
V_L		

2. Determine the phase angle of the circuit by using the following trigonometric methods:
   (a) Phase angle $(\phi) = \text{arc tangent } \frac{X_T}{R} =$ ______.
   (b) Phase angle $(\phi) = \text{arc tangent } \frac{V_X}{V_R} =$ ______.
   (c) Phase angle $(\phi) = \text{arc tangent } \frac{var_T}{W} =$ ______.

3. Determine the power factor of the circuit as:

   $$\text{Power factor (pf)} = \frac{\text{true power}}{\text{apparent power}} = \_\_\_\_\_\_.$$

4. Determine the cosine of the phase angle. This value should equal the power factor of the circuit.

   Cosine $\phi =$ ______.

5. How does the circuit of this experiment show power factor correction?

   ____

   ____

   ____

6. As total reactive power increases, the phase angle ______.

5-4

Parallel Resistive-Inductive (RL) Circuits

Objective

Another category of basic ac circuits is the parallel type. Parallel circuits are employed for many industrial control applications. The basic calculations necessary to study parallel ac circuits are somewhat different from those of series circuits. Since the impedance (Z) of a parallel circuit is less than the individual values of resistance, inductive reactance, or capacitive reactance, the impedance relationships must be modified. Also, a vectorial relationship exists between the currents that flow in the branches of the circuit.

In this laboratory activity, you will observe the impedance and current relationships of a parallel resistive-inductive (RL) circuit with ac applied to it.

Equipment

Ac voltage source.
Resistor—1000 Ω.
Inductor—8 H.
Electronic multifunction meter.

Procedure

1. Construct the parallel RL circuit shown in Fig. 5-4A.

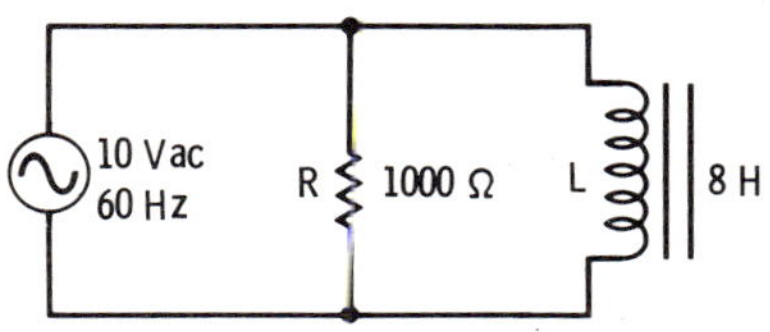

Fig. 5-4A. Parallel RL circuit.

2. Calculate the following:
   (a) Inductive reactance (X_L) = $2\pi fL$ =

   _______ ohms.

   (b) Current through the resistor (I_R) = $\frac{V}{R}$ =

   _______ amperes.

   (c) Current through the inductor (I_L) = $\frac{V}{X_L}$ =

   _______ amperes.

   (d) Total current (I_T) = $\sqrt{I_R^2 + I_L^2}$ =

   _______ amperes.

   (e) Impedance (Z) = $\frac{V}{I_T}$ = _______ ohms.

3. Complete the following current triangle for this circuit using the values from Step 2:

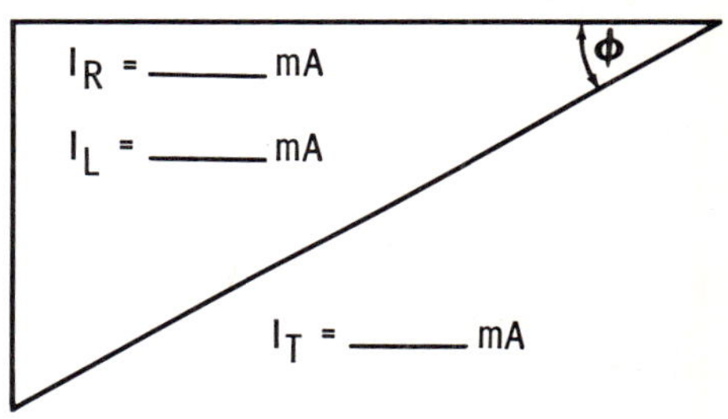

4. With a meter, measure the following values:
   (a) Current through the resistor (I_R) =

   _______ amperes.

   (b) Current through the inductor (I_L) =

   _______ amperes.

   (c) Total current (I_T) =

   _______ amperes.

5. Since the impedance is less than the resistance or inductive reactance of the circuit, an *admittance* diagram must be used. In this diagram, admittance (Y) = 1/Z, conductance (G) = 1/R, and inductive susceptance (B_L) = $1/X_L$. The unit of measurement is the mho or seimens (S).

6. Calculate the values necessary to complete the admittance triangle:

   (a) Admittance (Y) = 1/Z = _______ mho.

   (b) Conductance (G) = 1/R = _______ mho.

   (c) Inductive susceptance (B_L) = $1/X_L$ =

   _______ mho.

7. Complete the following admittance triangle for the circuit:

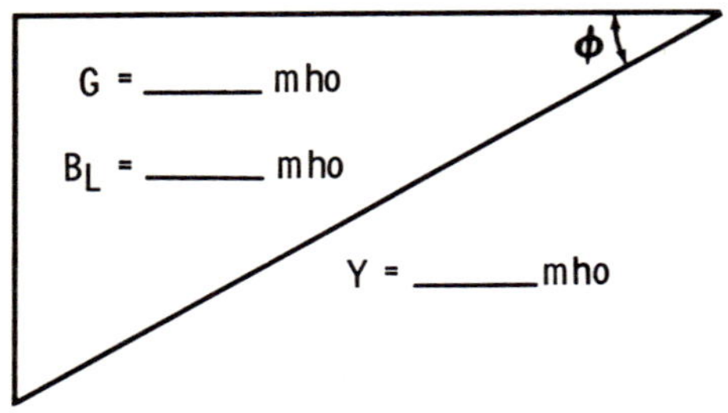

8. Calculate the power converted in the circuit:
   (a) True power (watts) $= I_R^2 \times R =$ ______ watts.
   (b) Apparent power (voltamperes) $= I_T^2 \times Z =$ ______ voltamperes.
   (c) Reactive power (var) $= I_L^2 \times X_L =$ ______ var.
9. Complete the following power triangle for the circuit:

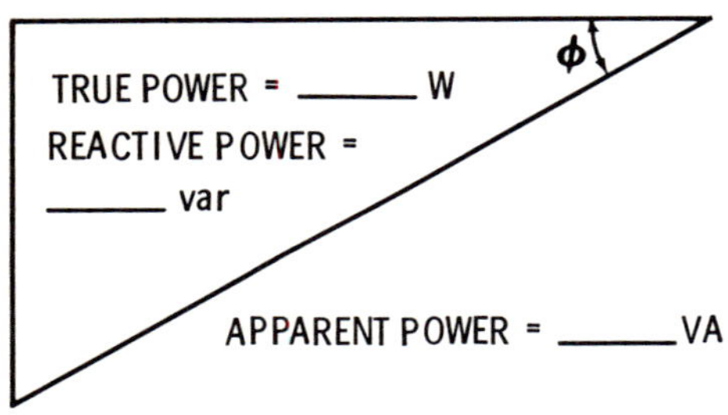

10. This concludes the experiment.

Analysis

1. Compare the calculated and measured values of the following:

Value	Calculated	Measured
I_R		
I_L		
I_T		

2. Determine the phase angle of the circuit by the following methods:

   (a) Phase angle (ϕ) = arc cosine $\frac{I_R}{I_T}$ = ______.

   (b) Phase angle (ϕ) = arc sine $\frac{\text{var}}{\text{VA}}$ = ______.
3. Determine the power factor of this circuit:

   Power factor (pf) $= \frac{\text{true power}}{\text{apparent power}} =$ ______.
4. Compare the power factor of this circuit with the value of cosine ϕ.
5. Define the following terms associated with parallel ac circuits:
   (a) Admittance.

   ______

   ______

   (b) Conductance.

   ______

   ______

   (c) Inductive susceptance.

   ______

   ______

   (d) Capacitive susceptance.

   ______

   ______
6. What are some sources of error which account for some difference between calculated and measured values in this activity?

   ______

   ______

   ______

Parallel Resistive-Capacitive (RC) Circuits

Objective

In this laboratory activity, you will observe the characteristics of a parallel RC circuit with ac applied. This type of circuit is similar to the RL circuit; however, the effect of capacitance is opposite that of inductance. Similar vectorial relationships exist in this circuit.

Equipment

Ac voltage source.
Resistor—1000 Ω.
Capacitor—1.0 μF.
Electronic multifunction meter.

Procedure

1. Construct the parallel RC circuit shown in Fig. 5-5A.

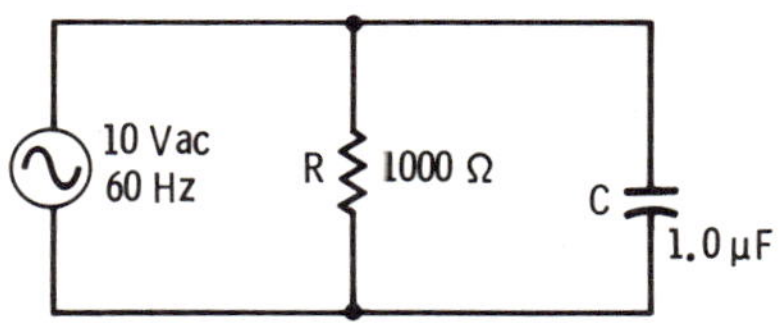

Fig. 5-5A. Parallel RC circuit.

2. Calculate the following:
   (a) Capacitive reactance $(X_C) = \frac{1}{2\pi fC} =$ ______ ohms.
   (b) Current through the resistor $(I_R) = \frac{V}{R} =$ ______ amperes.
   (c) Current through the capacitor $(I_C) = \frac{V}{X_C} =$ ______ amperes.
   (d) Total current $(I_T) = \sqrt{I_R^2 + I_C^2} =$ ______ amperes.
   (e) Impedance $(Z) = \frac{V}{I_T} =$ ______ ohms.

3. Complete the following current triangle for this circuit using the values from Step 2:

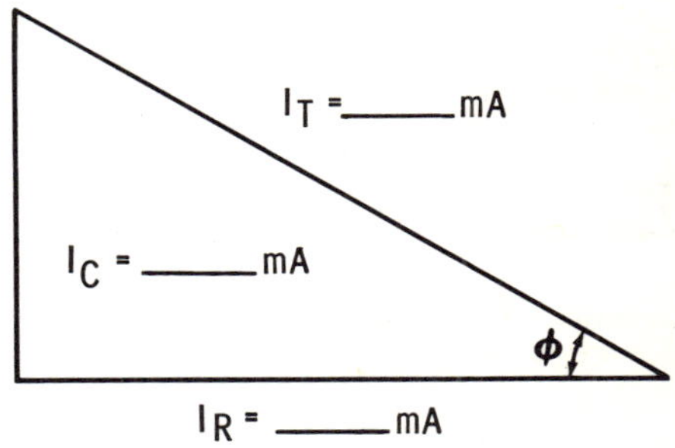

4. Measure the following ac current values:
   (a) Current through the resistor $(I_R) =$ ______ amperes.
   (b) Current through the capacitor $(I_C) =$ ______ amperes.
   (c) Total current $(I_T) =$ ______ amperes.

5. Calculate the following values:
   (a) Admittance $(Y) = 1/Z =$ ______ mho.
   (b) Conductance $(G) = 1/R =$ ______ mho.
   (c) Capacitive susceptance $(B_C) = 1/X_C =$ ______ mho.

6. Complete the following admittance triangle for the circuit:

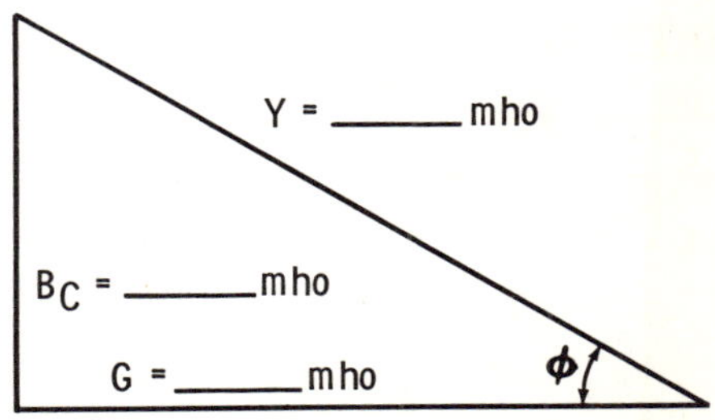

7. Calculate the power converted in the circuit:
   (a) True power $= I_R^2 \times R =$ ______ watts.

(b) Apparent power $= I_T^2 \times Z =$

_______ voltamperes.

(c) Reactive power $= I_C^2 \times X_C =$ _______ var.

8. Complete the following power triangle for the circuit:

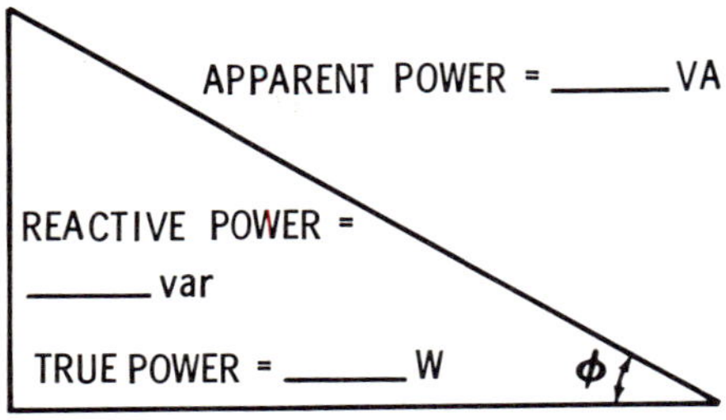

9. This concludes the experiment.

Analysis

1. Compare the calculated and measured values of the following:

Value	Calculated	Measured
I_R		
I_C		
I_T		

2. Determine the phase angle (ϕ) of the circuit using the following method:

(a) Phase angle (ϕ) $=$ arc sine $\frac{I_C}{I_T} =$ _______.

(b) Phase angle (ϕ) $=$ arc tangent $\frac{\text{var}}{\text{W}} =$ _______.

3. Determine the value of:

(a) Cosine $\phi =$ _______.

(b) Power factor $= \frac{\text{true power}}{\text{apparent power}} =$ _______.

4. If an 8-henry inductor was connected in parallel with the circuit of this activity, compute the following:

(a) $X_L = 2\pi fL =$ _______ ohms.

(b) $I_L = \frac{V}{X_L} =$ _______ amperes.

(c) $I_X = I_L - I_C =$ _______ amperes.

(d) $I_T = \sqrt{I_R^2 + I_X^2} =$ _______ amperes.

(e) $Z = \frac{V}{I_T} =$ _______ ohms.

(f) $\phi =$ arc sine $\frac{I_X}{I_T} =$ _______.

(g) pf $=$ cosine $\phi =$ _______.

5. How do the values of power factor for the original circuit and that of the preceding differ? Why?

____

____

____

6. How do the values of total current differ in the two circuits? Why?

____

____

____

5-6

Three-Phase Balanced Loads

Objective

An important consideration in three-phase power distribution is the type of loads connected to the power source. Three-phase motors are examples of balanced three-phase loads. The same value of impedance is connected across each phase of the power source. The balanced load is most desirable from a power economics point of view.

In this laboratory activity, you will construct a balanced three-phase load using three equal value resistors. This resistive load will be supplied by a low-voltage, three-phase source. The following three-phase power relationships should be noted:

(a) Power per phase $(P_p) = V_p \times I_p$.
(b) Total power $(P_T) = P_p \times 3 = V_L \times I_L \times 1.73$.
(c) For the delta system: $V_L = V_p$ and $I_L = I_p \times 1.73$.
(d) For the wye system: $V_L = V_p \times 1.73$ and $I_L = I_p$.

Equipment

Multifunction meter.
Variable three-phase power source.
Ac current meter.
Resistors (3)—1000 Ω, 10 W.

Procedure

1. Construct the three-phase, delta-connected, balanced resistive load circuit shown in Fig. 5-6A.
2. Apply 10 volts to the circuit from the three-phase source.
3. Using the multifunction meter and ac current meter, measure phase voltage (V_p), line voltage (V_L), phase current (I_p) and line current (I_L) for each phase. Record these values in Table 5-6A. Calculate and record the power per phase (P_p) and total power (P_T) values associated with each phase.
4. Complete the same measurements and calculations with 20, then 30 volts applied. Record these values in Table 5-6A.
5. Construct the wye-connected, balanced resistive load circuit shown in Fig. 5-6B.

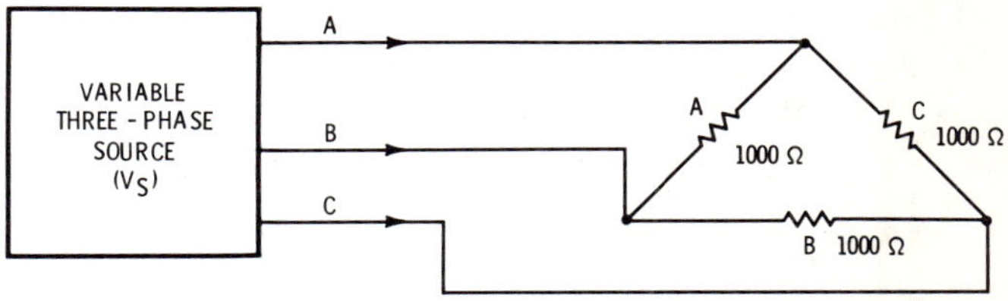

Fig. 5-6A. Three-phase, delta-connected, balanced resistive load circuit.

6. Perform the same series of measurements as those for the delta load. Record these values in Table 5-6B.
7. This concludes the experiment.

Analysis

1. From the data of Table 5-6A, compare:
   (a) $V_{p(A)}$ and $V_{L(A)}$ with 10 volts applied.

   ____

   ____

   (b) $I_{p(A)}$ and $I_{L(A)}$ with 20 volts applied.

   ____

   ____

   (c) $P_{p(A)}$ and $P_{p(B)}$ with 30 volts applied.

   ____

   ____

   (d) P_T with 10 volts applied and P_T with 30 volts applied.

   ____

   ____

   ____

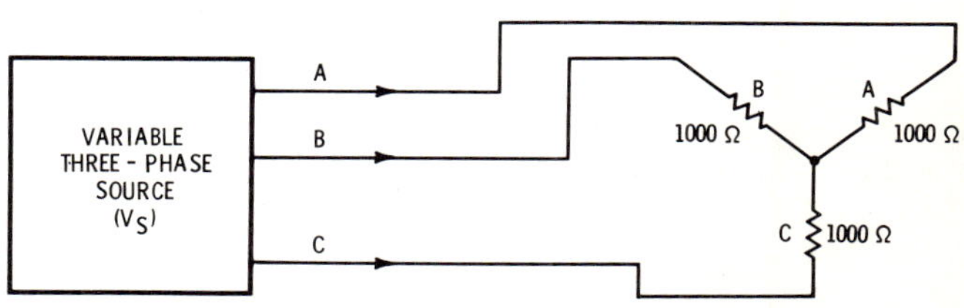

Fig. 5-6B. Three-phase, wye-connected, balanced resistive load circuit.

Table 5-6A. Delta Load Values

V_S	$V_{p(A)}$	$V_{p(B)}$	$V_{p(C)}$	$V_{L(A)}$	$V_{L(B)}$	$V_{L(C)}$	$I_{p(A)}$	$I_{p(B)}$	$I_{p(C)}$	$I_{L(A)}$	$I_{L(B)}$	$I_{L(C)}$	$P_{p(A)}$	$P_{p(B)}$	$P_{p(C)}$	P_T
10																
20																
30																

2. From the data of Table 5-6B, compare:
   (a) $V_{p(A)}$ and $V_{L(A)}$ with 10 volts applied.

   (b) $I_{p(A)}$ and $I_{L(A)}$ with 20 volts applied.

   (c) $P_{p(A)}$ and $P_{p(B)}$ with 30 volts applied.

   (d) P_T with 10 volts applied and P_T with 30 volts applied.

3. From the data of Tables 5-6A and B with 20 volts applied, compare:

   (a) $V_{p(A)}$ of Table 5-6A and $V_{p(A)}$ of Table 5-6B.

   (b) $I_{L(B)}$ of Table 5-6A and $I_{L(B)}$ of Table 5-6B.

   (c) $P_{p(C)}$ of Table 5-6A and $P_{p(C)}$ of Table 5-6B.

   (d) P_T of Table 5-6A and P_T of Table 5-6B.

Table 5-6B. Wye Load Values

V_S	$V_{p(A)}$	$V_{p(B)}$	$V_{p(C)}$	$V_{L(A)}$	$V_{L(B)}$	$V_{L(C)}$	$I_{p(A)}$	$I_{p(B)}$	$I_{p(C)}$	$I_{L(A)}$	$I_{L(B)}$	$I_{L(C)}$	$P_{p(A)}$	$P_{p(B)}$	$P_{p(C)}$	P_T
10																
20																
30																

5-7

Three-Phase Unbalanced Loads

Objective

In this laboratory activity, you will analyze three-phase circuits which represent unbalanced resistive loads. When the impedances connected across each phase of a three-phase system are unequal, an unbalanced situation exists. This condition is ordinarily encountered in three-phase systems when single-phase loads are supplied from three-phase lines. Most efficient operation of a three-phase system is obtained when the system is balanced. In unbalanced systems, total three-phase power (P_T), is calculated by adding the individual powers per phase (P_p). All other three-phase relationships for wye and delta systems remain the same.

Equipment

Multifunction meter.
Variable three-phase power source.
Ac current meter.
Resistors—1000 Ω, 10 W; 5000 Ω, 10 W; 10,000 Ω, 10 W.

Procedure

1. Construct the three-phase, delta-connected, unbalanced resistive load circuit shown in Fig. 5-7A.

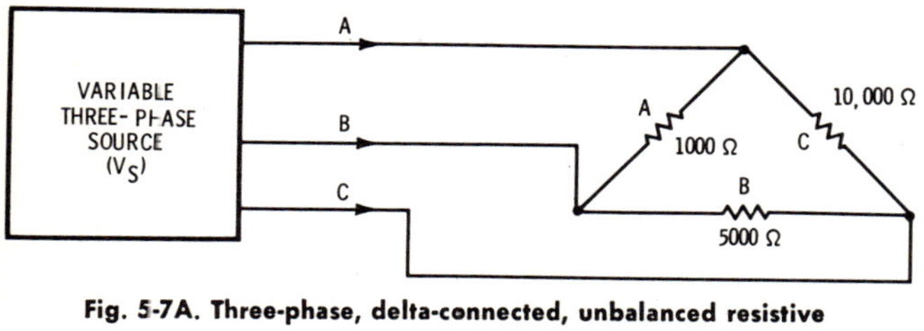

Fig. 5-7A. Three-phase, delta-connected, unbalanced resistive load circuit.

2. Apply 10 volts to the circuit from the three-phase power source.
3. Using the multifunction meter and ac current meter, measure phase voltage (V_p), line voltage (V_L), phase current (I_p), and line current (I_L) for each phase. Calculate the power per phase (P_p) and total power (P_T) values associated with each phase. Record each of these values in Table 5-7A.
4. Complete the same measurements and calculations for 20, then 30 volts applied. Record these values in Table 5-7A.
5. Construct the wye-connected, unbalanced resistive load circuit shown in Fig. 5-7B.

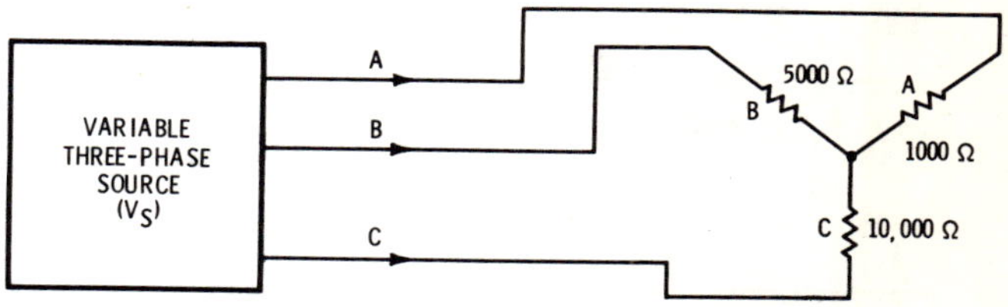

Fig. 5-7B. Three-phase, wye-connected, unbalanced resistive load circuit.

6. Perform the same measurements as those for the delta load. Record these values in Table 5-7B.
7. This concludes the experiment.

Analysis

1. From the data of Table 5-7A, compare:
   (a) $V_{p(A)}$ and $V_{L(A)}$ with 10 volts applied.

   ____

   ____

   (b) $I_{p(A)}$ and $I_{L(A)}$ with 20 volts applied.

   ____

   ____

   (c) $P_{p(A)}$ and $P_{p(B)}$ with 30 volts applied.

   ____

   ____

   (d) P_T with 10 volts applied and P_T with 30 volts applied.

   ____

   ____

Table 5-7A. Delta Load Values

V_S	$V_{p(A)}$	$V_{p(B)}$	$V_{p(C)}$	$V_{L(A)}$	$V_{L(B)}$	$V_{L(C)}$	$I_{p(A)}$	$I_{p(B)}$	$I_{p(C)}$	$I_{L(A)}$	$I_{L(B)}$	$I_{L(C)}$	$P_{p(A)}$	$P_{p(B)}$	$P_{p(C)}$	P_T
10																
20																
30																

2. From the data of Table 5-7B, compare:
   (a) $V_{p(A)}$ and $V_{L(A)}$ with 10 volts applied.

   (b) $I_{p(A)}$ and $I_{L(A)}$ with 20 volts applied.

   (c) $P_{p(A)}$ and $P_{p(B)}$ with 30 volts applied.

   (d) P_T with 10 volts applied and P_T with 30 volts applied.

3. From the data of Tables 5-7A and B with 20 volts applied, compare:
   (a) $V_{p(A)}$ of Table 5-7A and $V_{p(A)}$ of Table 5-7B.

   (b) $I_{L(A)}$ of Table 5-7A and $I_{L(A)}$ of Table 5-7B.

   (c) $P_{p(C)}$ of Table 5-7A and $P_{p(C)}$ of Table 5-7B.

   (d) P_T of Table 5-7A and P_T of Table 5-7B.

Table 5-7B. Wye Load Values

V_S	$V_{p(A)}$	$V_{p(B)}$	$V_{p(C)}$	$V_{L(A)}$	$V_{L(B)}$	$V_{L(C)}$	$I_{p(A)}$	$I_{p(B)}$	$I_{p(C)}$	$I_{L(A)}$	$I_{L(B)}$	$I_{L(C)}$	$P_{p(A)}$	$P_{p(B)}$	$P_{p(C)}$	P_T
10																
20																
30																

5-8

Electrical Lighting

Objective

Electrical lighting loads constitute one of the major types of loads used today. Lighting loads convert electrical energy into light energy. In this laboratory activity, you will be able to observe some of the characteristics of incandescent, fluorescent, and vapor lamps. These three types of lighting are the most commonly used.

Equipment

Incandescent lamp—40 W.
Fluorescent lamp.
Fluorescent ballast.
Fluorescent starter.
Vapor lamp—neon.
Vapor lamp—argon.
Variable ac power source.
Multifunction meter.

Procedure

1. Construct the circuit shown in Fig. 5-8A for investigating incandescent lighting. Use a clear 40-watt lamp mounted in a socket.

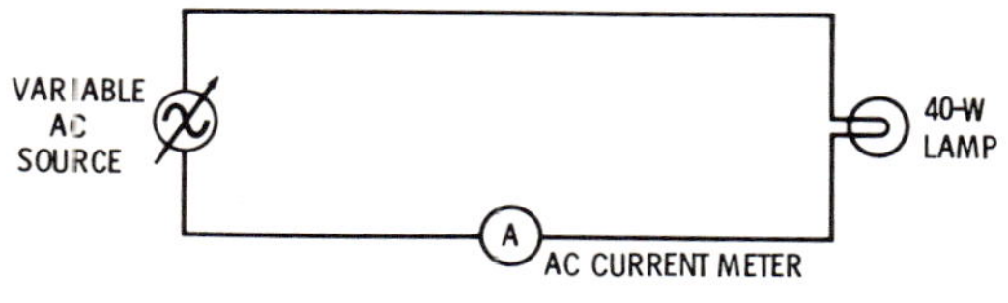

Fig. 5-8A. Incandescent lighting circuit.

2. Beginning with zero volts ac, increase the ac voltage while observing the lamp filament. Note that when the filament changes from an orange color to a white glow, the filament appears to change diameter. This is the point of incandescence.
3. Measure and record the minimum voltage required to produce incandescence.

   Voltage = ______ volts ac.
4. Apply 120 volts ac to the circuit. Measure and record the ac current flowing through the lamp.

   Current = ______ amperes ac.
5. Remove the lamp from the socket and allow some time for it to cool. Measure the resistance of the lamp filament with the meter.

   Resistance = ______ ohms.
6. Place a neon vapor lamp in the socket. Determine the minimum voltage that causes the lamp to glow. This voltage is called the *ionization* or *firing point.*

   Ionization voltage = ______ volts ac.
7. Observe the color of the glow.

   Color = ____________.
8. Apply 120 volts ac to the lamp. Measure and record the ac current indicated on the meter.

   Current = ______ amperes ac.
9. Remove the lamp from the socket and measure its resistance.

   Resistance = ______ ohms.
10. Place an argon lamp in the socket and repeat Steps 6–9. Argon lamp data:

    (a) Ionization voltage = ______ volts ac.

    (b) Color of glow = ____________.

    (c) Current = ______ amperes ac.

    (d) Resistance = ______ ohms.
11. Construct the fluorescent lighting circuit shown in Fig. 5-8B.

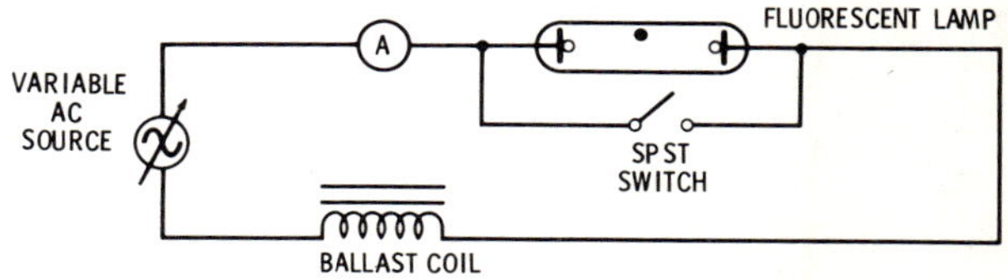

Fig. 5-8B. Fluorescent lighting circuit.

12. Adjust the variable ac source to 120 volts ac to apply power to the circuit. Momentarily turn on the spst switch. This should cause the filaments at each end of the fluorescent tube to glow. Now, turn this switch off. What occurs?

____

____

13. With the light functioning, reduce the ac voltage source until the light stops glowing. Measure and record the voltage where this condition occurs.

    Light stops glowing at ______ volts ac.

14. Exchange the spst switch with an automatic fluorescent starter. Apply 120 volts ac to the circuit and check the circuit for operation. Measure and record the ac current.

    Current = ______ amperes ac.

15. Remove the starter from the circuit. Again, apply 120 volts ac to the circuit. What occurs?

____

____

Analysis

1. What is the purpose of the ballast coil in a fluorescent lamp circuit? Starter?

____

____

____

2. What are some applications of vapor lamps?

____

____

____

3. What is meant by the term *ionization point* in conjunction with vapor lamps? Compare these values for the neon and argon lamps.

____

____

____

4. Explain the relationship between resistance and wattage of an incandescent lamp.

____

____

____

5. Explain how the fluorescent lamp operates.

____

____

____

6. From Step 4, calculate the resistance of the lamp as: $R = V/I$. Compare this value to the measured resistance in Step 5. Why do they differ?

____

____

____

7. What are the colors produced by the following vapor lamps?

   (a) Neon: ______.

   (b) Argon: ______.

   (c) Mercury: ______.

8. Compare the values of current drawn by the incandescent (Step 4), neon vapor (Step 8), argon vapor (Step 10), and fluorescent (Step 14) lamps investigated in this experiment. Be sure to note the wattage rating of each lamp.

____

____

____

____

5-9

Electrical Heating

Objective

Heat-producing loads constitute one of the major types of industrial loads. There are several types of heating loads used in industry. The three basic electrical heating methods are resistive, inductive, and capacitive. Electrical welders and plant heating systems convert electrical energy into heat energy.

In this laboratory activity, resistive and inductive heating methods will be observed. You should keep in mind that one of the most important applications of electricity today is to produce heat.

Equipment

Conical heating coil—660 W.
Lamp socket.
Electromagnetic coil.
Ac ammeter.
Multifunction meter.
Variable ac source.
Soft iron core.

Procedure

1. Construct the simple resistive heating circuit shown in Fig. 5-9A. Do not apply power now.

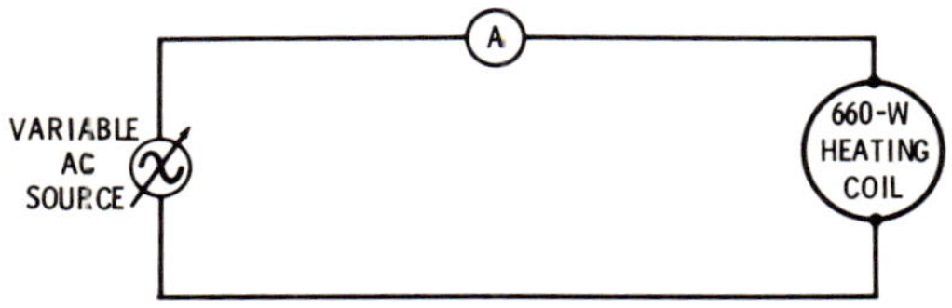

Fig. 5-9A. Resistive heating circuit.

2. Measure and record the cold resistance of the 660-watt heating unit.

   Resistance = ______ ohms.
3. Apply 120 volts ac to the circuit and allow the coil to heat for about two minutes. Quickly remove the power source and measure the resistance of the heater.

   Hot resistance = ______ ohms.
4. Adjust the variable ac power source to produce the necessary current readings listed in Table 5-9A. Measure and record the applied voltage for each line of Table 5-9A. After each voltage is recorded, disconnect the heating coil quickly and measure its hot resistance.
5. Connect the inductive heating circuit shown in Fig. 5-9B.

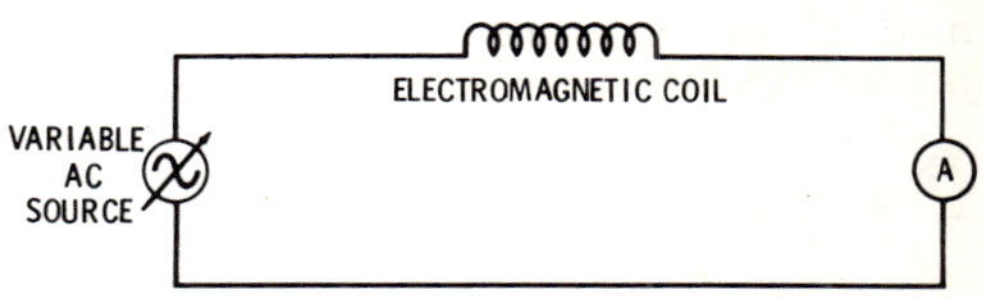

Fig. 5-9B. Inductive heating circuit.

6. Place a round, soft-iron core inside the electromagnetic coil. With 5 volts ac applied, record the current indicated. (This voltage may have to be changed, depending on the resistance of the coil.)

   Current = ______ amperes ac.
7. Remove the iron core and record the current.

   Current = ______ amperes ac.

Table 5-9A. Resistive Heating Values

AC Current	Voltage	Power (V × A)	Hot Resistance	Calculated Heat (calories or joules)*
1.0 ampere				
1.5 amperes				
2.0 amperes				
2.5 amperes				
3.0 amperes				

* 0.24 calorie = 1.0 watt-second = 1.0 joule. (Use 1.0 second for the time.)

8. Adjust the variable ac power source to obtain the current values listed in Table 5-9B. Touch the iron core after each current adjustment to determine the extent to which it has heated. This illustrates the inductive heating effect.
9. This concludes the experiment.

Table 5-9B. Inductive Heating Values

AC Current	Voltage	Apparent Power (V × A)	Heat Produced*
0.5 ampere			
1.0 ampere			
1.5 amperes			
2.0 amperes			

* Indicate amount of heat, such as warm, hot, very hot.

Analysis

1. What causes a resistive heating element to produce heat?

2. From the data in Steps 2 and 3, how does the resistance of the heating device change with temperature?

3. How are the wattage and the resistance of a resistive heating device related?

4. Compare the current values of Steps 6 and 7. Are they different? Why?

5. From the data of Table 5-9B, as current increases, what happens to the amount of heat produced?

6. How does inductive heating occur?

7. How does capacitive heating occur?

8. Why does a higher frequency of applied ac improve the heating effect of inductive or capacitive heating elements?

5-10

Ignitron Characteristics

Objective

Ignitrons are commonly used with high-current welders in industry. Because of their use in high-current applications, it is important to have an understanding of their operation. In most cases, it is not feasible to use ignitrons in a laboratory situation.

In this activity, you will become familiar with the construction and functional characteristics of an ignitron through the identification of its parts and by analyzing the operation of two ignitron circuits.

Procedure

1. Identify and label the parts of the ignitron shown in Fig. 5-10A.
2. Identify and label the parts of the ignitron schematic symbol shown in Fig. 5-10B.

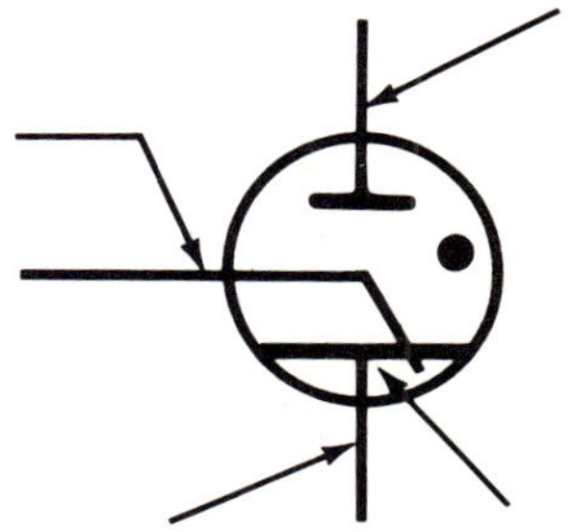

Fig. 5-10B. Ignitron schematic symbol.

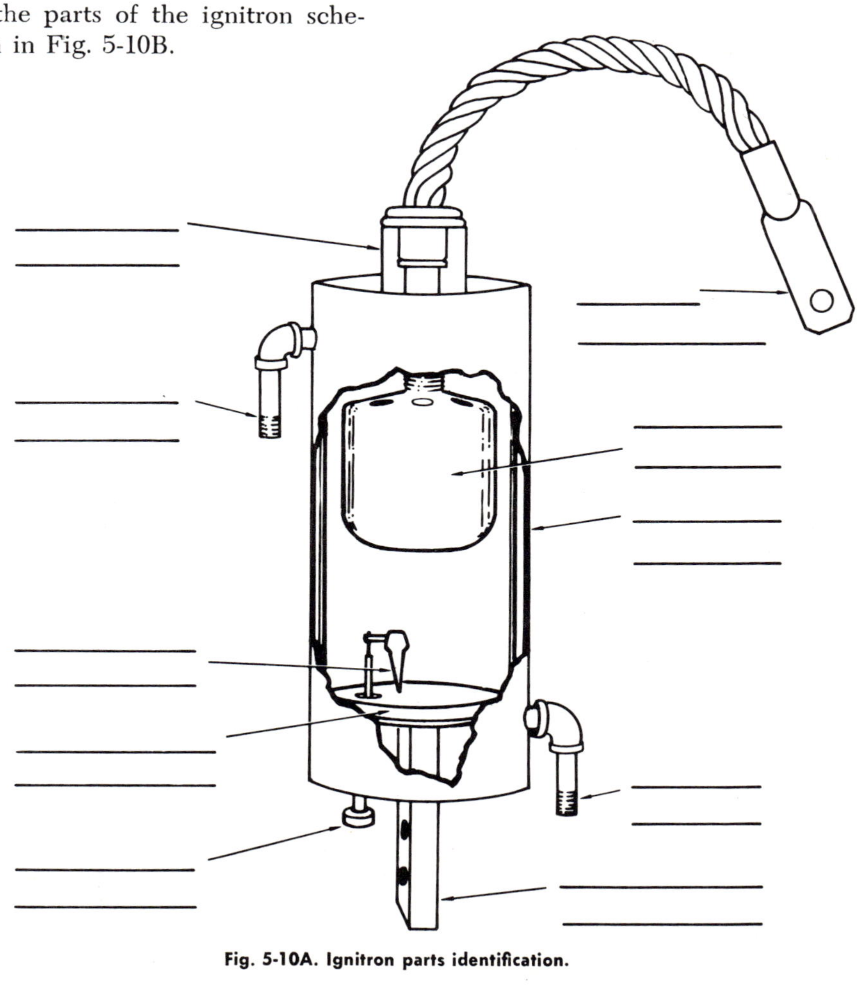

Fig. 5-10A. Ignitron parts identification.

3. Analyze the operation of the two ignitron circuits shown in Fig. 5-10C. Explain the purpose of each component used in the circuits.

____

____

____

____

____

____

____

____

____

____

____

____

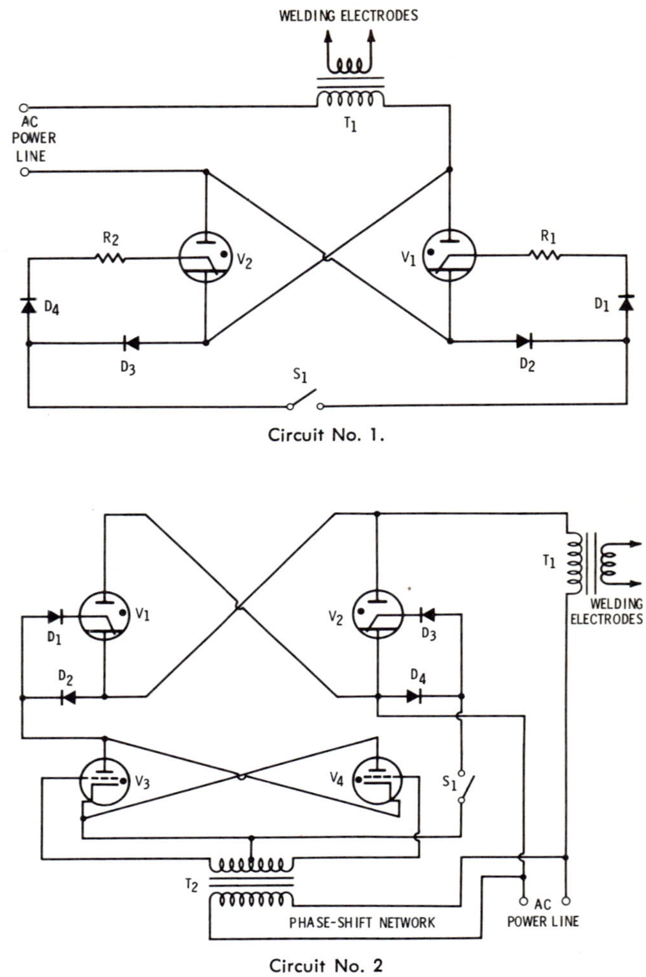

Fig. 5-10C. Ignitron welding circuits.

5-11

Series-Wound DC Motor Analysis

Objective

The series-wound dc motor is constructed of low-resistance field coils connected in series with the armature coils. The field coils have a few turns of large diameter wire. This arrangement permits a current flow that is the same throughout the circuit. Like all motors, the series-wound machine slows down with increased loads and speeds up with decreased loads. Its speed, like other dc motors, is inversely proportional to its magnetic field strength. The counterelectromotive force, or cemf, which is generated when the armature turns through the magnetic field, is directly related to speed. As a result, these three factors, speed, cemf, and field strength, are extremely important characteristics in the operation of a series-wound dc motor.

In this laboratory activity, you will be able to analyze the operation of a series-wound dc motor. This type of motor is commonly designed so that it will operate from either a dc or ac voltage source. Such machines are called *universal* motors. This activity is set up in such a way that you can analyze: (1) a commercially manufactured dc series motor, such as a small universal motor used in an electric drill or an automobile starter, or (2) a dc series motor that has been constructed using a rotating machinery unit.

Equipment

Multifunction meter.
Series-wound dc motor.
Dc power supply.
Tachometer.

Procedure

1. The connection diagram for a series-wound dc motor is shown in Fig. 5-11A.

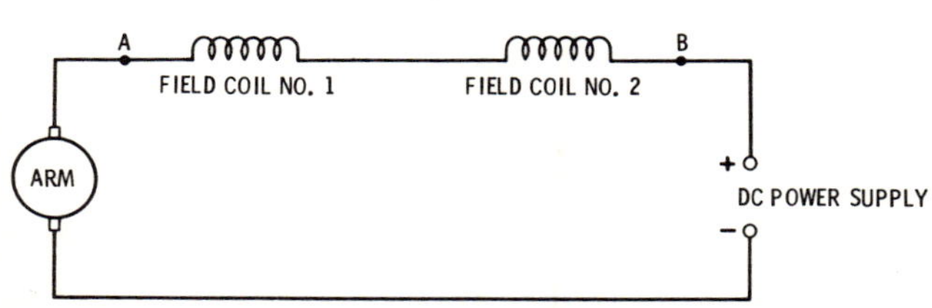

Fig. 5-11A. Connection diagram for a series-wound dc motor.

2. Obtain a dc series motor or construct one using the rotating machinery unit.
3. Measure the resistance of the armature. This will be a low resistance, so use the lowest resistance range on the meter.

   R_A = ______ ohms.
4. Measure the resistance of the field coils (point A to point B on the diagram). Make sure the coils are isolated from other parts of the circuit. It may be necessary to remove a wire at point A.

   R_F = ______ ohms.
5. Prepare a meter to measure direct current and connect it in series with the motor circuit.
6. Gradually increase the dc voltage applied to the machine until you reach its rated voltage. Observe the effect on current and speed.
7. With rated voltage applied, turn the power supply off and then back on momentarily. Observe the effect on current as speed increases toward its maximum.
8. Carefully apply a load on the motor shaft while observing the effect on current and speed.
9. With a tachometer, measure the motor speed at its rated voltage.

   Speed = ______ r/min.
10. Record the current flow indicated at the rated voltage of the machine.

    Current = ______ amperes.
11. Turn the power supply off and use the necessary procedure to *reverse* the direction of rotation of the motor.
12. Turn on the power supply and measure the speed and current of this machine with its direction of rotation reversed.

    Speed = ______ r/min.

    Current = ______ amperes.
13. Try to apply a heavy enough load to stall the motor. Very quickly record the maximum current the motor draws.

    Maximum current = ______ amperes.

14. Allow the motor to run for a few seconds to cool off.
15. This concludes the experiment.

Analysis

1. What effect would having more than two field coils have on the operation of this motor?

2. What effect does increased voltage have on motor speed? Why? (See Step 6).

3. In Step 7, what happened to motor speed and current as load was increased? Why?

4. How is the direction of rotation of the dc series motor reversed?

5. Was there any change of motor speed or current when the direction of rotation was reversed? Should there have been?

6. Show in the following graph generally how a speed-versus-load curve would look.

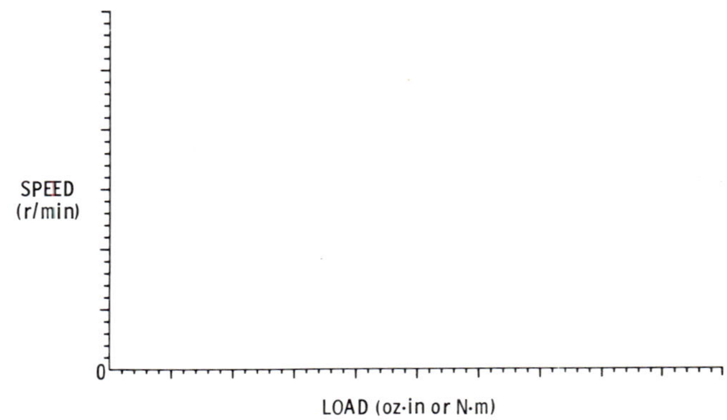

7. Calculate the input power to this motor at its rated voltage.

   Watts input = V × I = ______ watts.

8. Convert the watts input (W) to horsepower (hp).

$$hp^* = \frac{W}{746} = \_\_\_\_\_\_ \text{ hp.}$$

*This value is not the output horsepower rating of the machine. The output horsepower would be somewhat lower due to energy losses of the machine.

5-12

Shunt-Wound DC Motor Analysis

Objective

The construction of dc motors is based on the method of connecting the field windings and armature coils together. In the shunt-wound dc motor, for example, the field coils are connected in parallel with the armature. This is achieved by making the shunt field a fairly high resistance. Accordingly, it is wound of many turns of fine wire. The armature, however, offers less resistance and permits a larger current flow. Its construction is of a larger wire with fewer turns of wire.

In this laboratory activity, you will be able to analyze either: (1) a commercially manufactured dc shunt motor, or (2) a dc shunt motor that has been constructed in the laboratory using a rotating machinery unit. You should observe the general construction and operational characteristics of the shunt-wound dc motor.

Equipment

Multifunction meter.
Shunt-wound dc motor.
Variable dc power supply.
Field rheostat (0–500 Ω, high wattage).
Armature rheostat (0–25 Ω, high wattage).
Tachometer.

Procedure

Section A. Shunt-Wound DC Motor Characteristics

In this section of the laboratory activity, you will observe the operational characteristics of a shunt-wound dc motor. You will observe the effect of load on the speed, counterelectromotive force (cemf), current, and torque of this machine.

1. The connection diagram for a shunt-wound dc motor is shown in Fig. 5-12A.
2. Measure the resistance of the field coils (point A to point B on the diagram). You will have to remove a wire at point A. It will be necessary to use the lowest resistance range on your meter.

   R_F = ______ ohms.

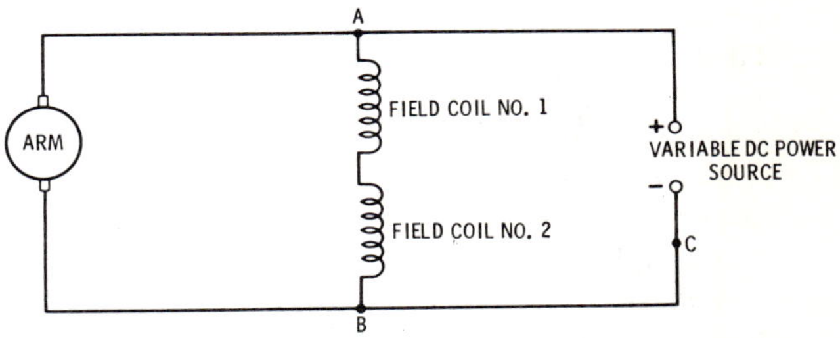

Fig. 5-12A. Connection diagram for a shunt-wound dc motor.

3. Measure the resistance of the armature circuit (across the brushes). Make sure the power supply is not connected to the armature, and the wire at point A is still removed. This will isolate the armature resistance from other circuits.

   R_A = ______ ohms.
4. Prepare a meter to measure a high value of dc current. Put the current meter in series with the power supply (at point C on the diagram). This meter will measure the total current drawn by the motor.
5. Record the rated voltage value of this motor.

   Rated voltage = ______ volts dc.
6. Turn on the power supply and slowly increase the voltage to the rated voltage of the motor. Observe the effect on current and speed of the motor.
7. Record the total current (I_T) drawn by the motor at its rated voltage.

   I_T = ______ amperes dc.
8. With a tachometer, measure and record the speed of the motor.

   Speed = ______ r/min.
9. Carefully apply a load to the motor shaft by putting pressure on the shaft, which will slow its speed. Observe the effect that increased load has on speed and total current.
10. Increase the load as much as possible, or until rotation stops or slows down appreciably. Quickly record the maximum current drawn by the motor.

    Maximum current = ______ amperes dc.

11. Turn off the power supply and perform the necessary procedure to reverse the direction of rotation of the motor.
12. With the direction of rotation reversed, increase the applied dc voltage to the rated voltage value of the motor.
13. Measure and record the current and speed.

    Current = ______ amperes dc.

    Speed = ______ r/min.
14. Turn off the power supply.

Section B. Speed Control of Shunt-Wound DC Motors

In this section, you will observe two methods of controlling the speed of a dc shunt motor. These methods are very simple. The first method utilizes a rheostat in series with the armature circuit. The second method employs a rheostat in series with the field circuit. By measuring current flow through each rheostat and calculating power dissipation, you should be able to decide which speed control method is most desirable.

1. Modify the dc shunt motor as shown in Fig. 5-12B. Be sure the wattage ratings of the rheostats are high enough.
2. The rheostats should be connected in series with the armature and field circuits as shown in the diagram. Adjust each rheostat to the zero resistance position.
3. Adjust the variable dc power supply from zero volts to the rated voltage of the motor to check it for operation.
4. Turn off the power supply and connect a dc current meter in series with the armature circuit at point A on the diagram.
5. With rated voltage applied, turn on the motor and observe the maximum starting current indicated on the meter.

    Maximum starting current = ______ amperes dc.
6. Also, measure and record the running current of the motor with no load applied.

    Running current = ______ amperes dc.
7. Adjust the armature rheostat from zero toward its maximum resistance while observing the speed of the motor. As armature resistance increases, speed

    ____________. With a tachometer, measure the range of speed.

    Speed range = ______ to ______ r/min.

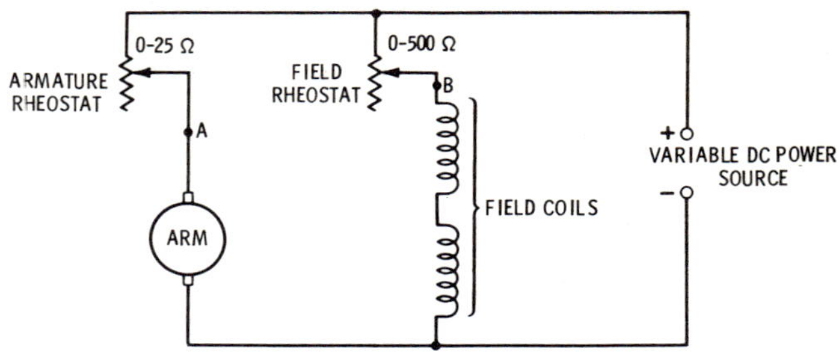

Fig. 5-12B. Shunt-wound dc motor speed-control test circuit.

8. Adjust the armature rheostat back to zero resistance. Carefully apply a load to the motor shaft. Measure and record the maximum armature current as load is increased.

    Maximum armature current = ______ amperes dc.
9. Turn off the power supply. Now, remove the armature current meter and connect the meter in series with the field circuit at point B on the diagram.
10. With rated voltage applied, turn on the motor and observe the maximum field current during starting.

    Maximum field current = ______ amperes dc.
11. Record the field current indicated while the motor is running with no load applied.

    Field current = ______ amperes dc.
12. Adjust the field rheostat from zero toward its maximum resistance while observing the speed of the motor. As field resistance increases, speed

    ____________. With a tachometer, measure the range of speed.

    Speed range = ______ to ______ r/min.
13. Adjust the field rheostat back to zero resistance. Carefully apply a load to the shaft of the motor. Measure and record the maximum field current as load is increased.

    Maximum field current = ______ amperes dc.
14. Turn off the power supply. This concludes the experiment.

Analysis

1. How does the field resistance of a dc shunt motor compare to that of a dc series motor?

    ____

    ____

    ____

2. How does increased load on a dc shunt motor affect:
   (a) Speed of the motor?

   (b) Cemf developed by the armature?

   (c) Total current drawn by the motor?

   (d) Torque developed by the motor?

3. In Section A, Step 11, how was the direction of rotation of the motor reversed?

4. Did reversing the direction of rotation have any effect on total current and speed (see Section A, Steps 7, 8, and 13)? Why?

5. How do the starting current and running current drawn by the armature compare (see Section B, Steps 5 and 6)? Why?

6. Using the maximum armature current value from Section B, Step 5, and the measured armature resistance, determine the minimum power rating of an armature rheostat.

   $P_A = I_A^2 \times R_A =$ ______ watts.

7. In Section B, Step 7, as armature resistance changes, why does speed change as it does?

8. How do starting current and running current drawn by the field compare (see Section B, Steps 10 and 11)? Why?

9. In Section B, Step 12, as field resistance changes, why does speed change as it does?

10. Using the maximum field current obtained in Section B, Step 13, and the measured field resistance, determine the minimum power rating of a field rheostat for speed control.

    $P_F = I_F^2 \times R_F =$ ______ watts.

11. Which type of speed control (armature or field) is most desirable? Why?

Split-Phase AC Motor Analysis

Objective

Single-phase ac motors are generally classified according to their method of starting. A very common method employed for starting single-phase motors is phase splitting. In this method, different inductive reactances are used to shift the phase of ac in such a way that it will produce rotation. Once rotation is started and the rotor reaches 75% of its running speed, the starting circuit is no longer needed because the motor runs on the induction principle. As a result, most starting circuits are removed after being in the circuit for only a few seconds. Generally, a switch mechanism that operates on centrifugal force, mounted on the rotor shaft, is used to remove the starting circuit from the power source.

In this laboratory activity, you will be able to analyze a split-phase ac motor. These motors are used for many applications that do not require a high starting torque. The procedure of this activity is organized in such a way that you may either analyze a commercially manufactured split-phase ac motor or one that has been constructed using a rotating machinery unit.

Equipment

Multifunction meter.
Split-phase ac motor.
Ac power source.
Tachometer.
Ac ammeter.

Procedure

1. The connection diagram for a split-phase ac motor is shown in Fig. 5-13A.
2. There are several modifications of this circuit; however, the run windings and start windings are always connected in parallel, with the centrifugal switch in series with the start windings. Dual-voltage motors (120/240 volts, for example) have two sets of run windings. They are connected in parallel for low voltage operation and in series for high voltage operation.
3. Observe the arrangement of the start and run windings around the stator assembly. How many

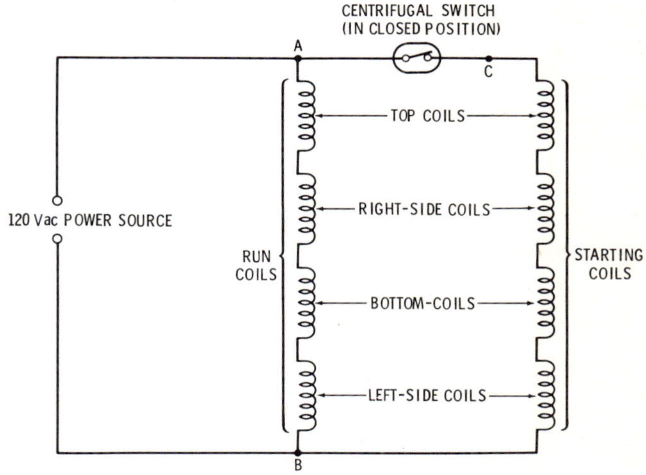

Fig. 5-13A. Connection diagram for a split-phase ac motor.

run-winding coil sets (poles) does your motor have? How many starting-coil sets?

_______ run-coil sets.

_______ starting-coil sets.

4. Trace the external connecting wires of the motor to their origin within the stator. Momentarily disconnect the necessary wires to make the following resistance measurements. Use a low resistance range of the meter. Measure and record run-winding resistance (point A to point B) and start-winding resistance (point C to point B).

R_{run} = _______ ohms.

R_{start} = _______ ohms.

5. Make sure all wires are properly connected so that the motor will be operational. Connect an ac ammeter in series with the 120-V ac line.
6. Apply the rated voltage to the motor from the power source. Observe and record the starting current drawn by the motor when it is initially turned on.

Starting current = _______ amperes ac.

7. Now, record the current flow indicated when the motor is running with no load applied.

Running current = _______ amperes ac.

8. With a tachometer, measure and record the operating speed of the motor.

   Speed = ______ r/min.
9. Carefully apply a load to the motor shaft. Observe the effect on speed and current.
10. Turn off the power and perform the necessary procedure to reverse the direction of rotation of the motor.
11. Turn on the power and test the motor for operation in the reverse direction. Measure and record the line current and speed.

    Current = ______ amperes ac.

    Speed = ______ r/min.

12. This concludes the experiment.

Analysis

1. How is starting torque developed in a split-phase ac motor?

   ____

   ____

   ____

2. In the space below, draw diagrams of a dual-voltage (120/240-volt) motor showing how its windings are connected, first for low voltage operation and then for high voltage operation.

3. Based on the number of running poles observed in Step 3 of the experiment, determine the synchronous speed of the motor you analyzed.

$$\text{Synchronous Speed (r/min)} = \frac{\text{ac frequency} \times 120}{\text{number of poles}} = \_\_\_\_\_\_.$$

4. Using the operating speed measured in Step 8, determine the slip percentage of the motor with no load applied.

$$\%\ \text{Slip} = \frac{\text{synchronous speed} - \text{operating speed}}{\text{synchronous speed}} \times 100 = \_\_\_\_\_\_.$$

5. The rotor frequency (f_r) affects both the starting and running characteristics of an induction motor. Determine the rotor frequency of the motor you analyzed as follows:

$$f_r\ (\text{Hz}) = S \times f_s = \_\_\_\_\_\_,$$

   where S is the slip, expressed as a decimal, and f_s is the stator frequency, expressed in hertz.
6. When looking at the stator windings of a split-phase ac motor, how do you know which sets of coils are the run windings and which are the start windings?

   ____

   ____

   ____

7. Complete the following statements by filling in the word *high* or *low:* Comparatively, the start windings of an ac split-phase motor have ________ resistance and ________ inductive reactance, while the run windings have ________ resistance and ________ inductive reactance.

8. How do the values of starting and running current compare? Why?

____

____

____

9. In Step 9, what was the effect of increased load on speed and current?

____

____

10. In Step 10, how was the direction of rotation of the motor reversed?

____

____

11. In Step 11, when rotation was reversed, what was the effect on current and speed as compared to Steps 7 and 8? Why?

____

____

____

5-14

Single-Phase AC, Capacitor-Start Motor Analysis

Objective

Single-phase induction motors are not self-starting without some auxiliary method used to provide initial movement of the rotor. One method is the use of a starting winding. Single-phase induction motors that employ auxiliary windings for starting may have a capacitance connected in series. A capacitor-start induction motor is made to operate as though a two-phase ac source supplied the power to produce the rotating magnetic field. A capacitor is employed to provide initial starting torque and to develop a strong, uniform magnetic field.

In this laboratory activity, you will be able to analyze the operational characteristics of either a commercially manufactured capacitor motor or one that has been constructed in the laboratory. You should be able to see the effect the capacitor has on the starting torque of the motor.

Equipment

Multifunction meter.
Single-phase ac power source.
Ac current meter.
Capacitor-start motor.
Tachometer.

Procedure

1. A connection diagram for a single-phase ac, capacitor-start induction motor is shown in Fig. 5-14A.

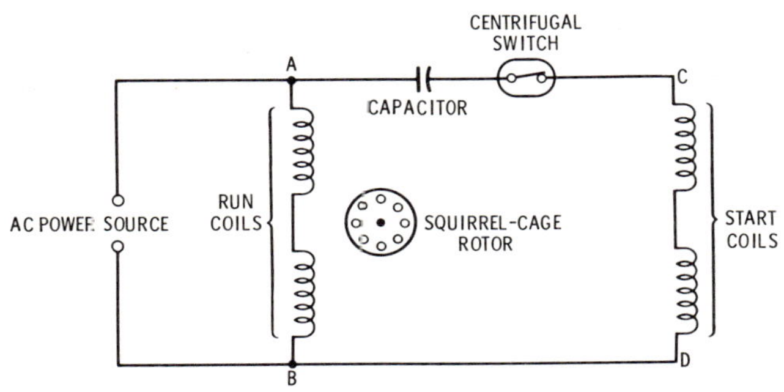

Fig. 5-14A. Connection diagram for a single-phase ac, capacitor-start induction motor.

2. Count the number of stator-coil sets (poles) that the motor you are analyzing has in the run and start circuits.

   Number of running-coil sets = ______.

   Number of starting-coil sets = ______.
3. With a meter, measure the resistance of the run windings (point A to point B on the diagram). Temporarily disconnect the run windings to make this measurement.

   R_{run} = ______ ohms.
4. Now, measure the resistance of the start windings (point C to point D on the diagram).

   R_{start} = ______ ohms.
5. Connect an ac current meter in series with one of the power lines. Apply rated voltage to the motor and observe the initial current drawn by the motor.

   Starting current = ______ amperes.
6. Now, record the running current of the motor.

   Running current = ______ amperes.
7. Carefully apply a load to the shaft of the motor while observing the effect on speed and current.
8. With a tachometer, measure the speed of rotation.

   Speed = ______ r/min.
9. Turn off the motor and place a wire from one terminal of the capacitor to the other. This will bypass the capacitor so that it no longer functions in the starting circuit.
10. Turn the power on and observe the operation of the motor. Then, remove the wire from the capacitor.
11. If possible, connect another capacitor of the same or similar value in parallel with the starting capacitor. Record the total capacitance in the start circuit.

    Total capacitance = ______ μF.

12. With the additional capacitance in the circuit, turn on the power and observe the operation of the motor. Record the starting current.

    Starting current = ______ amperes.

13. With the motor in its original circuit configuration, perform the necessary procedure to reverse the direction of rotation. Measure and record the starting current, running current, and speed.

    Starting current = ______ amperes.

    Running current = ______ amperes.

    Speed = ______ r/min.

14. This concludes the experiment.

Analysis

1. Why does the addition of a capacitor in the start windings of an induction motor improve its starting torque?

2. Describe the following types of capacitor motors:
   (a) Capacitor-start induction motor.

   (b) Capacitor-start, capacitor-run motor.

   (c) Permanent-capacitor motor.

3. From the data of Step 2, compare the number of starting and running coil sets of the motor.

4. From the data of Steps 3 and 4, compare the resistances of the start and run windings of the motor.

5. From the data of Steps 5 and 6, compare the starting current and running current values.

6. How does increased load affect the speed and current of the motor (see Step 7)?

7. In Step 10, what happened when the capacitor was bypassed?

8. What does the addition of more capacitance in the start circuit have on starting current of the motor?

9. How is the direction of rotation of the capacitor-start motor reversed?

10. What are some applications of capacitor motors?

5-15

Shaded-Pole AC Motor Analysis

Objective

A unique method of making a single-phase ac motor self-starting is by a process called *pole-shading.* The shading poles are contained within the field poles of the machine. Generally, about one-third of the field pole is enclosed by the shading coil. When the magnetic flux around the pole starts to build up it cuts across the shading coil. This induces a voltage into the shading coil that produces a current flow. The area inside of the shading coil then has a reverse polarity compared with the remaining pole face. As a result, there is a shifting effect created across the pole face from the unshaded to the shaded side. The shifting effect is significant enough to cause the rotor to turn. This turning action then makes the motor self-starting. The starting torque of the shaded-pole motor is very small, with some of the largest ratings of this type of motor around 1/20 horsepower.

The shaded-pole motor is mainly used for very small torque applications such as fan or blower motors. In this laboratory activity, you will be able to analyze either a commercially manufactured, ac shaded-pole motor or one that has been constructed in the lab.

Equipment

Multifunction meter.
Shaded-pole ac motor.
Ac power source.
Tachometer.
Ac current meter.

Procedure

1. The connection diagram for a shaded-pole ac motor is shown in Fig. 5-15A.
2. Measure the resistance of the stator windings.

   Winding resistance = ______ ohms.
3. Connect an ac current meter in series with the power line. Apply rated voltage to the motor and observe the initial current it draws when turned on.

   Starting current = ______ amperes.

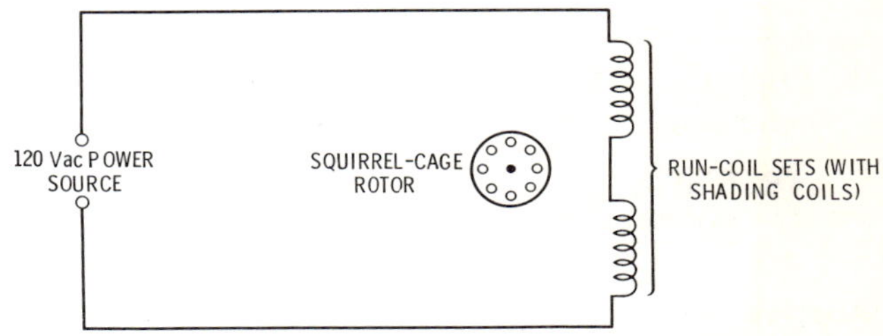

Fig. 5-15A. Connection diagram for a shaded-pole ac motor.

4. Now, record the current drawn by the motor with no load applied.

   Running current = ______ amperes.
5. Carefully apply a load to the shaft of the motor while observing the effect on speed and current.
6. With a tachometer, measure the speed of rotation.

   Speed = ______ r/min.
   How many poles or coil sets does the motor have?

   ______ poles.
7. This concludes the experiment.

Analysis

1. Explain how the shaded-pole induction motor produces starting rotation.

   ____
   ____
   ____
2. What are some applications of shaded-pole motors?

   ____
   ____
   ____
3. How is the synchronous speed of all ac induction motors determined?

   ____
   ____
   ____

4. What is the synchronous speed of the motor you analyzed?

____

5. What is meant by the term *slip* as applied to ac induction motors?

____

____

____

6. From the data of Step 6, determine the slip percentage of the motor you analyzed.

$$\% \text{ Slip} = \frac{\text{synchronous speed} - \text{operating speed}}{\text{synchronous speed}} \times 100 = \_\_\_\_\_\_.$$

7. How do the starting and running currents of the shaded-pole motor compare? Why?

____

____

____

8. What is the effect of increased load on motor speed and current?

____

____

____

9. How could the direction of rotation of a shaded-pole motor be reversed?

____

____

____

10. What would happen if one of the shading coils became open-circuited?

____

____

____

5-16

Three-Phase AC Induction Motor Analysis

Objective

The three-phase ac induction motor has the simplest construction features of all ac motors. Due to the inherent characteristics of three-phase current, no auxiliary starting equipment is required for this motor. This allows the motor to operate efficiently without utilizing start windings, centrifugal switches, shaded poles, or any other auxiliary parts. Three-phase induction motors have very good starting and running torque. They are more economical to use (per horsepower) than other ac motors.

Since three-phase current is not distributed to homes, most applications of three-phase ac induction motors are in industry. Seldom are ratings of less than 5 horsepower used, and 500–1000 horsepower motors of this type are commonly used. You could say that the three-phase induction motor is the "workhorse" of industry, since it is used for so many applications.

In this laboratory activity, you will be able to analyze a commercially manufactured, three-phase ac induction motor or one that has been constructed in the laboratory. You should observe its simple construction features and its high-torque, low-current operating characteristics.

Equipment

Multifunction meter.
Three-phase ac induction motor.
Ac current meter.
Three-phase ac power source.
Tachometer.

Procedure

1. One connection diagram for a three-phase ac induction motor is shown in Fig. 5-16A. This illustrates a wye-connected stator with three stator-coil sets. The stator coils could also be delta-connected. The motor with six stator coils, as shown in the diagram, is the minimum number a three-phase motor can have.
2. Count the number of stator-coil sets (poles) the motor you are analyzing has.

   Number of stator-coil sets = ______.
3. With a meter, measure the resistance of one of the coil sets. Be sure it is temporarily disconnected in order to obtain an accurate resistance measurement.

   Winding resistance = ______ ohms.

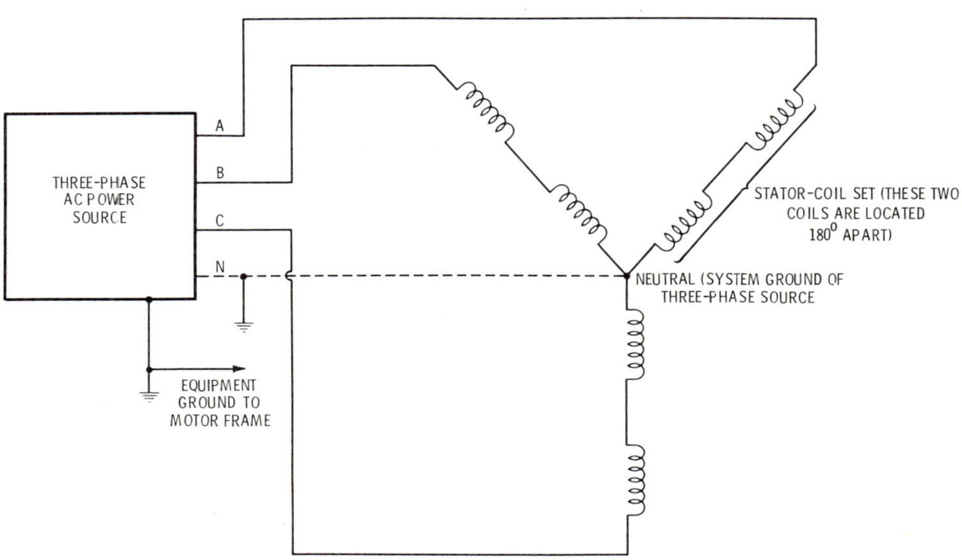

Fig. 5-16A. Connection diagram for a three-phase ac induction motor with wye-connected stator coils.

Table 5-16A. Starting and Running Current Values (Wye)

Phase	Starting Current	Running Current
A		
B		
C		

4. Assure that the stator windings are connected in a wye configuration. Measure the winding resistance across a pair of three-phase lines.

   Winding resistance = ______ ohms.
5. Connect an ac current meter in series with one of the three-phase lines (refer to this phase as phase A) to measure starting and running current. Turn on the three-phase power source and apply the rated voltage. Measure the values of starting and running current. Record in Table 5-16A.
6. Now, connect the ac ammeter in series with phase B, then phase C, to measure starting and running currents. Record these values in Table 5-16A.
7. With a tachometer, measure the speed of rotation.

   Speed = ______ r/min.
8. Turn off the power supply and reverse the direction of rotation of the motor. Measure the speed of rotation.

   Speed = ______ r/min.
9. Carefully apply a load to the shaft of the motor. Observe the effect on speed and current. Turn off the power source when finished. Now, reconnect the stator windings in a delta configuration as shown in Fig. 5-16B.
10. Measure the starting and running currents for the delta-connected stator windings. Record these values in Table 5-16B.
11. Measure the speed of rotation with a tachometer.

    Speed = ______ r/min.
12. Turn off the power source and reverse the direction of rotation of the motor. Measure its speed.

Table 5-16B. Starting and Running Current Values (Delta)

Phase	Starting Current	Running Current
A		
B		
C		

Turn off the power source when finished.

Speed = ______ r/min.

13. This concludes the experiment.

Analysis

1. How does the three-phase ac induction motor develop a starting torque?

   ____

   ____

   ____
2. Using the data from Step 2, determine the synchronous speed of this motor.

   $$\text{Synchronous speed (r/min)} = \frac{\text{ac frequency} \times 120}{\text{number of poles per phase}} = \_\_\_\_\_\_.$$
3. Compare the winding resistance measurements of Steps 3 and 4. Why are they different?

   ____

   ____

   ____
4. Compare the starting and running current values recorded in Table 5-16A.

   ____

   ____

   ____

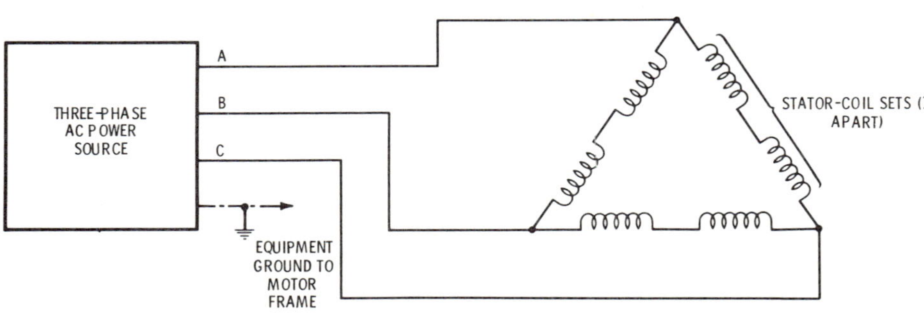

Fig. 5-16B. Connection diagram for a three-phase ac induction motor with delta-connected stator coils.

5. Using the operating speed measured in Step 7, determine the slip percentage of this motor.

$$\% \text{ Slip} = \frac{\text{synchronous speed} - \text{operating speed}}{\text{synchronous speed}} \times 100 = \_\_\_\_.$$

6. How is the direction of rotation of a three-phase ac motor reversed?

7. Compare the current values recorded in Table 5-16A for a wye-connected stator with those of Table 5-16B for a delta-connected stator.

8. How does the r/min value for the wye connection compare with that of the delta connection? Why?

9. What effect does increased load have on the operation of a three-phase ac induction motor?

5-17

Three-Phase AC Synchronous Motor Analysis

Objective

Three-phase ac synchronous motors operate at a constant speed regardless of the mechanical load applied. In addition to the constant speed characteristic, this type of motor also has the ability to operate at leading or lagging power factors. By changing the dc excitation voltage applied to the rotor, the power factor at which the motor operates will vary. When a synchronous motor operates at a leading power factor, it improves the overall power factor of the system on which it is operating. This operating mode produces a capacitive effect on the system; therefore, the machine is referred to as a *synchronous capacitor*.

In this laboratory activity, you will have an opportunity to analyze the operational characteristics of a commercially manufactured, three-phase ac synchronous motor, or one that has been constructed in the lab using a rotating machinery unit. The construction of this motor is the same as that of a three-phase ac alternator. Therefore, a three-phase alternator, such as one used in an automobile, could be substituted for an operational three-phase ac synchronous motor for a laboratory analysis.

Equipment

Multifunction meter.
Three-phase ac synchronous motor.
Three-phase ac power source.
Variable dc power source.
Ac current meter.
Tachometer.

Procedure

1. A simplified connection diagram for a three-phase ac synchronous motor is shown in Fig. 5-17A. Normally, the stator coils are connected in a wye configuration.
2. Measure the following resistances across the power lines:

   $R_{A\text{-}B}$ = ______ ohms.

   $R_{B\text{-}C}$ = ______ ohms.

   $R_{A\text{-}C}$ = ______ ohms.

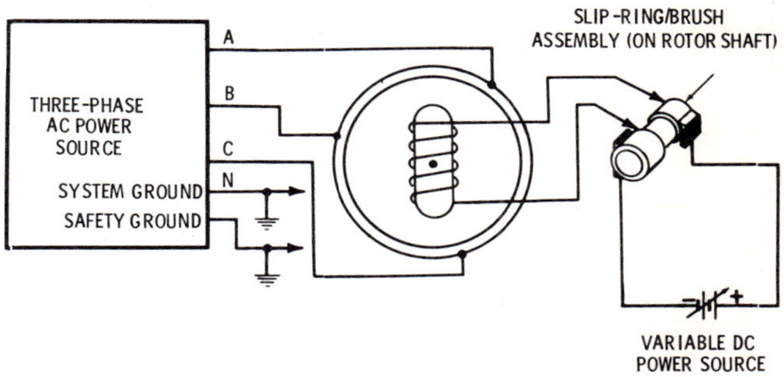

Fig. 5-17A. Connection diagram for a three-phase ac synchronous motor.

3. Connect the synchronous motor to a three-phase ac power source in order to supply the rated voltage to the stator coils of the motor.
4. Place an ac current meter in series with one of the power lines.
5. Connect the terminals from the slip-ring/brush assembly of the rotor to a variable dc power source. Adjust this dc voltage to zero.
6. Determine the number of stator-coil sets (poles) that your motor has.

   Stator-coil sets = ______.
7. Apply three-phase ac power to the machine. Does it start?

   ____
8. If the machine does not start, and if it does not have damper windings, it may be necessary to bring its speed up to near its synchronous speed by connecting another motor to its shaft to supply the necessary speed.
9. With the synchronous motor operational, record the ac current indicated on the meter.

   Current = ______ amperes ac.
10. Now, gradually increase the dc voltage applied to the rotor, while observing the effect on speed and ac current.
11. Turn the dc voltage down to zero once more. Connect a dc current meter in series with the power source.
12. Now measure the speed of rotation with a tachometer.

   Speed = ______ r/min.

Table 5-17A. Rotor Current Versus Stator Current

DC Voltage	Rotor Current (DC)	Stator Current (AC)
0		

13. If available, record the rated dc voltage of the machine you are analyzing.

    ______ volts dc at ______ power factor.
14. Vary the dc voltage applied to the rotor in increments of approximately 2 volts until about 50% above the rated dc voltage is reached. At each dc voltage value, record the dc current through the rotor and the ac current drawn by the stator. Record your data in Table 5-17A.
15. With the rated dc voltage applied to the rotor, measure the speed of rotation with a tachometer.

    Speed = ______ r/min.
16. Carefully apply a load to the shaft of the machine. Observe the effect of increased load on ac stator current, dc rotor current, and speed.
17. Turn the dc voltage back to zero volts dc. Now gradually increase the dc voltage until the rotor is synchronized. You should note an effect called "hunting," which is evident as a noise produced as the rotor becomes synchronized in step with the revolving stator field.
18. Turn off the dc, then the ac power source.
19. This concludes the experiment.

Analysis

1. Describe the basic construction of a three-phase synchronous motor.

    ______

    ______

    ______
2. What is the speed of the following three-phase synchronous motors?

    (a) 6 pole* = ______ r/min.

    (b) 12 pole* = ______ r/min.

    (c) 24 pole* = ______ r/min.
3. How does a three-phase synchronous motor differ from a three-phase induction motor?

    ______

    ______

    ______
4. What are some methods of starting three-phase synchronous motors?

    ______

    ______

    ______
5. What is pull-in torque? Pull-out torque?

    ______

    ______

    ______
6. When is a three-phase synchronous motor considered to be underexcited? Overexcited?

    ______

    ______

    ______
7. Verify your observation of the number of stator poles in Step 6 with the speed of rotation measured in Step 15:

    $$\text{Synchronous speed (r/min)} = \frac{\text{ac frequency} \times 120}{\text{number of poles/phase}} = \_\_\_\_\_\_.$$
8. What effect does increased rotor current have on the stator current of the synchronous motor?

    ______

    ______

    ______
9. What effect does increased load have on (see Step 16):

    (a) Speed?

    ______

    ______

    (b) Dc rotor current?

    ______

    ______

*Refers to total number of stator poles.

(c) Ac stator current?

____

____

10. What effect does increasing the dc rotor voltage (from zero) have on the speed of the motor?

____

____

____

11. With the data recorded in Table 5-17A, make a graph of dc rotor current versus ac stator current. Also, mark the point on the curve that corresponds to the highest power factor operation of the motor.

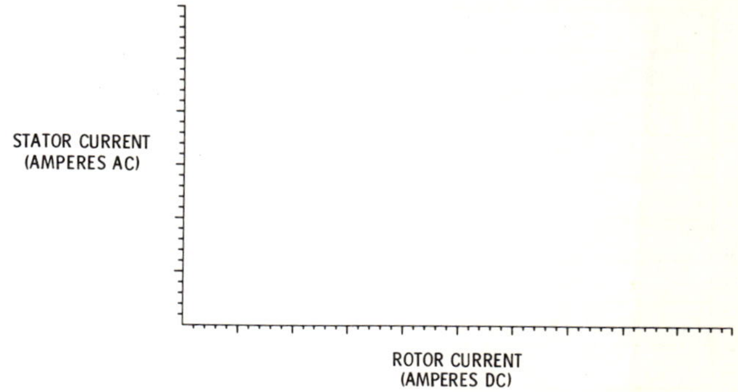

5-18

Dynamometer Analysis of Electric Motors

Objective

A very thorough analysis of the operation of any motor under varying load conditions can be made by using a dynamometer. The dynamometer unit should be used to analyze and compare the characteristics of many types of fractional horsepower motors. Most commonly, 1/4-, 1/3-, or 1/2-horsepower motors are used on laboratory dynamometer units.

In this laboratory activity, you should analyze as many motors as possible. There is no particular order in which they should be done. When you complete an analysis, you should submit to the instructor the following:

1. Motor nameplate data (Table 5-18A).
2. Motor analysis data (Table 5-18B).
3. Motor data graph (Fig. 5-18B) with the following performance curves:
   (a) Speed vs torque.
   (b) Speed vs horsepower.
   (c) Efficiency vs torque.
4. A written analysis in two parts:
   (a) General information about the operational characteristics of this type of motor.
   (b) Reference to the data obtained to discuss the characteristics of the motor. (Discuss each characteristic curve on the graph.)

Equipment

Dynamometer unit.
Wattmeter.
Ac ammeter.
Multifunction meter.
Graph paper (one sheet for each motor).
Calculator.

Procedure

1. Select a motor for analysis and record the nameplate data of the motor in Table 5-18A.
2. Secure the motor onto the dynamometer unit. Make sure the shaft couplings are of the proper type for the motor you are analyzing.

Table 5-18A. Motor Nameplate Data

Motor Type	
Manufacturing Co.	
Identification Number	
Model Number	
Frame Type	
Number of Phases (AC)	
Horsepower	
Cycles (AC)	
Speed (r/min)	
Voltage Rating	
Current Rating (Amperes)	
Thermal Protection	
Temperature Rating—°C	
Time Rating	
Other Information	

3. Connect the motor to the current and power meters as shown in Fig. 5-18A. Be sure to connect the wattmeter properly.
4. Adjust the load applied to the motor by the dynamometer through as many increments of torque as possible. Record the current and power for each value of torque in Table 5-18B.
5. Using the data recorded in Table 5-18B, and the rated voltage of the motor, calculate the horsepower, efficiency, and power factor for each value

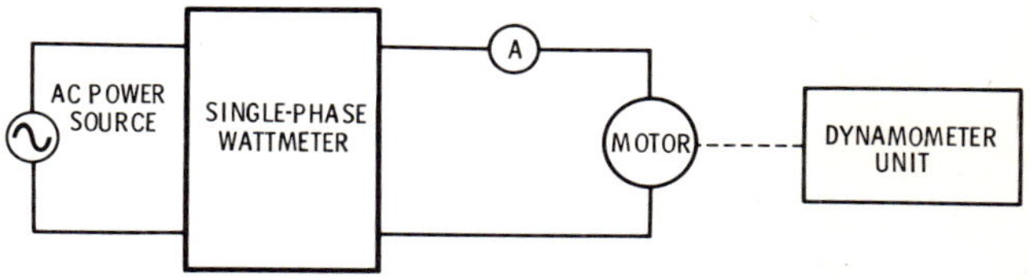

Fig. 5-18A. Connection diagram for dynamometer analysis of electric motors.

Table 5-18B. Motor Analysis Data

Torque (ft · lb)	Speed (r/min)	True Power (Watts)	Line Current (Amperes)	Horsepower*	% Efficiency†	Power Factor‡
1						
5						
10						
15						
20						
25						
etc.						

$$\text{* Horsepower} = \frac{\text{torque (ft} \cdot \text{lb)} \times \text{speed (r/min)}}{5250}$$

$$\text{† \% Efficiency} = \frac{\text{horsepower} \times 746}{\text{true power (watts)}}$$

$$\text{‡ Power Factor} = \frac{\text{true power (watts)}}{\text{apparent power (voltamperes)}}$$

of torque. Record these calculated values in the appropriate columns of Table 5-18B.

6. On a sheet of graph paper plot the performance curves for the motor data graph by using the values of your recorded data. Refer to the sample graph of Fig. 5-18B for selection of the proper horizontal and vertical axes.
7. Prepare your data and analysis to be subbmitted for evaluation. Try to do a neat and accurate job on your data and graph, and a detailed job on the analysis.
8. Perform the same procedure on as many motors as possible. It may be interesting to compare the curves obtained from several different types of motors.
9. This concludes the activity.

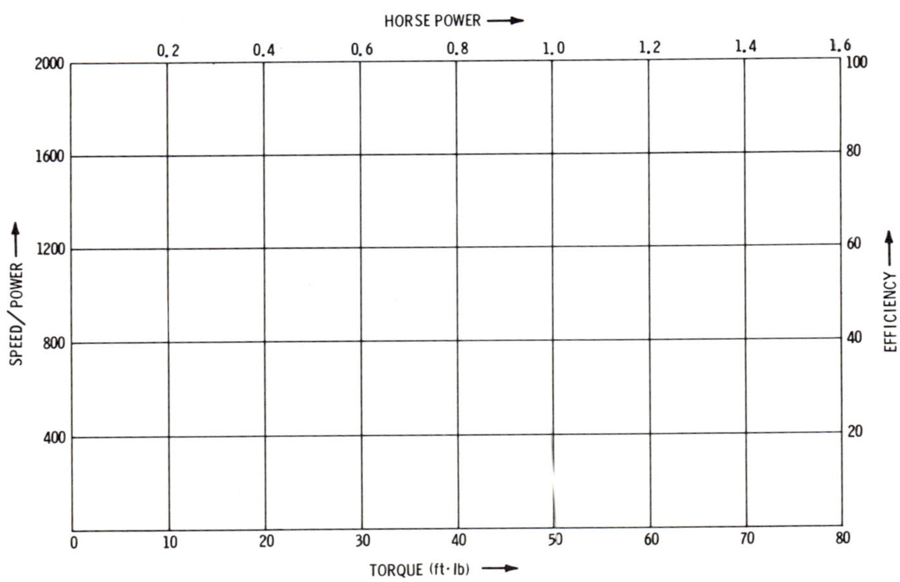

Fig. 5-18B. Motor data graph.

Motor Repair and Troubleshooting Procedure

Objective

Motor troubleshooting should be a systematic process by which malfunctions are traced and repaired. An electrical technician should be able to systematically locate and repair many motor malfunctions. For effective troubleshooting, you should have a practical understanding of the type of motor that is being repaired. With this basic knowledge of operation and some test equipment, you should be able to perform many repair jobs.

Some of the following instruments may be used to make various tests on the motor: (1) tachometer—speed test; (2) megger—insulation test; (3) multifunction meter—resistance, current, and voltage test; (4) growler—wound armature test; and (5) motor dynamometer analysis unit—speed vs torque test. You may need to check with the instructor on the use of some of this equipment as your troubleshooting progresses.

In this laboratory activity, you will be able to systematically test and repair a motor of any type. Many parts of the procedure listed below may not be applicable to your particular troubleshooting job. However, try to follow most of the listed steps and complete the analysis.

Equipment

Motor that does not function properly.
Multifunction meter.
Growler (optional).
Megger (optional).
Dynamometer (optional).
Tachometer (optional).

Procedure

1. Examine the external features of the motor you have selected to troubleshoot.

2. Find the motor nameplate and record the information called for below. If not available, indicate so in the blank.

Manufacturing Co. ______________________

Motor Type ______________________

Identification Number ______________________

Model Number ______________________

Frame Type ______________________

Number of Phases (AC) ______________________

Horsepower ______________________

Cycles (AC) ______________________

Speed (r/min) ______________________

Voltage Rating ______________________

Current Rating (Amperes) ______________________

Thermal Protection ______________________

Temperature Rating—°C ______________________

Time Rating ______________________

Other Information ______________________

____

____

____

____

3. Determine what type of motor you are working with, such as: ac induction, repulsion, shaded-pole, or capacitor-start. Consult with the instructor to determine the specific type. What is the specific type?

____

4. In the following space, draw a schematic diagram of the motor on which you are working. This will provide a general reference for later tests.

5. After the proper type of power source and voltage has been determined, apply power to the motor for a very brief period of time while attempting to detect the specific malfunction.

6. What malfunctions seem to exist in your motor? Describe them.

____

____

____

____

____

____

____

____

7. If available, consult a trouble chart for this type of motor in a motor operation book or some other reference. List some possible causes of the malfunction of your motor.

____

____

____

____

____

____

____

____

8. Determine some possible remedies to alleviate the malfunction. List some of the remedies.

____

____

____

____

____

____

____

____

9. If the motor must be disassembled, remove the end plates carefully. Be particularly careful with the connecting wires and bearings. Remove the rotor assembly at this time. Check the bearings.

10. If connecting wires must be removed, use tape labels or some other means to label the proper terminal connections.

11. You may need to make some resistance tests with the multifunction meter to determine the condition of the stator windings. (Example: Ac single-phase induction motor; start winding—20 ohms resistance; run winding—5 ohms resistance.) Of course, those resistances will vary. Record your resistance readings.

Winding resistances = ____________________

____

____

____

12. What can you conclude from the resistance test regarding the condition of your motor?

____

____

____

____

____

____

____

____

13. Now inspect all wiring connections, splices, and other mechanical connections for possible opens or shorts.
14. Reinspect the stator portion of the motor for excessive dirt, oil, or other damage.
15. Check the rotor for possible troubles that may be visible.
16. If the rotor is the squirrel-cage type, it may be in need of cleaning with fine sandpaper.
17. If a wound rotor is used, check the commutator for high mica insulation, roughness, irregular wear, or other damage. You may wish to dress the commutator on a metal lathe, clean it with sandpaper, or remove high mica with an undercutting tool.
18. Inspect rotor windings for damaged insulation. You may test the rotor windings for damaged insulation with a megger. You may test the rotor for shorts, grounds, or open circuits with a growler.
19. If the motor has brushes, check for excessive wear, oil, grease, chipping, or improper contact.
20. Perform any other tests you feel may be needed to locate the trouble. Consult with your instructor.
21. Perform the necessary procedures to repair the motor.
22. Consult with your instructor when you feel you have completed the troubleshooting procedure and have repaired the motor malfunction. Have him approve your procedures.

Instructor's Approval: ____________________

23. This should complete your troubleshooting job.

Analysis

1. Write a detailed description of the troubleshooting process you have used in this experiment. Explain how you determined the trouble for this particular type of motor. Also, explain how you remedied the trouble.

____

____

____

____

____

____

____

____

____

____

____

____

____

____

____

____

____

____

____

____

____

____

____

____

____

____

5-20

Motor Control—Dynamic Braking

Objective

When a motor is turned off, its shaft will continue to rotate for a short period of time. This continued rotation is undesirable for many industrial applications. Dynamic braking is a method used to bring a motor to a quick stop when power is turned off. Motors with wound armatures utilize a resistance connected across the armature as a dynamic braking method. When power is turned off, the resistance is connected across the armature. This causes the armature to act as a loaded generator, making the motor slow down immediately.

In this laboratory activity, you will use simple dynamic braking methods to stop a single-phase ac induction motor and a dc shunt motor.

Equipment

Single-phase ac induction motor (120 volts).
Dc shunt motor.
Switch—Double-pole, double-throw.
Resistor—10 Ω, 20 W.

Procedure

1. Construct the dynamic brake for a single-phase ac induction motor as shown in Fig. 5-20A.
2. Place the dpdt switch in the "run" position, applying 120 volts ac to the motor.
3. Open the switch (center position) and observe the time required for the motor shaft to stop rotating.

   Estimated time for motor to stop = ______ seconds.
4. Once again place the dpdt switch in the "run" position. Now, place the switch in the "brake" position while observing the time required to stop the motor.

   Estimated time for motor to stop = ______ seconds.
5. This concludes the dynamic braking procedure for a single-phase induction motor.
6. Construct the circuit shown in Fig. 5-20B to demonstrate dynamic braking of a dc shunt motor.

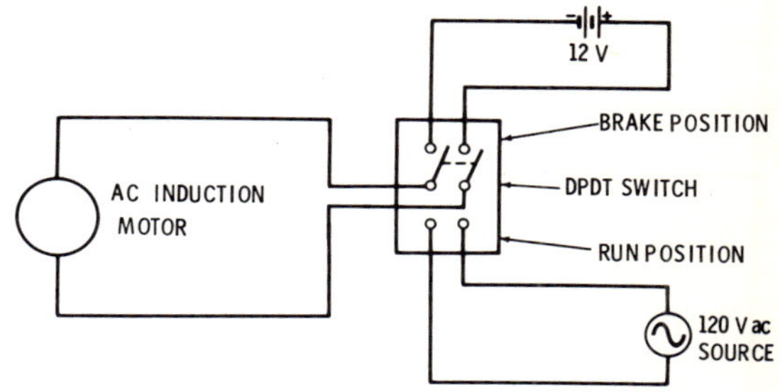

Fig. 5-20A. Dynamic braking circuit for a single-phase ac induction motor.

7. Apply rated voltage to the dc motor by placing the dpdt switch in the "run" position.
8. Open the switch (center position) and observe the time required for the motor shaft to stop rotating.

   Time for motor to stop = ______ seconds.
9. Again, place the switch in the "run" position. Now, place the switch in the "brake" position while observing the time required for the motor to stop.

   Time for motor to stop = ______ seconds.
10. This concludes the experiment.

Analysis

1. What are some situations where dynamic braking of motors is needed?

____

____

____

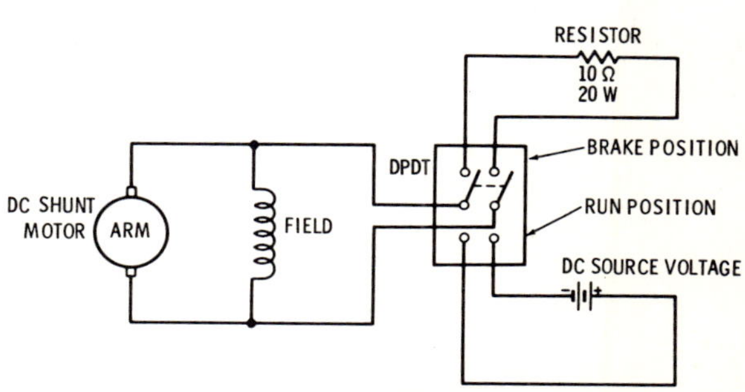

Fig. 5-20B. Dynamic braking circuit for a dc shunt motor.

2. Discuss the method used for dynamic braking of an induction motor.

____

____

____

3. Discuss the method used for dynamic braking of a dc motor.

____

____

____

4. Draw a schematic diagram of a switching system that could be used for dynamic braking of a series-wound dc motor.

5-21

Motor Control With a Magnetic Contactor

Objective

A very common method used to turn electric motors on or off is a magnetic contactor used in conjunction with a set of push-button switches. A magnetic contactor is a type of relay. It is energized by applying a voltage to its electromagnetic coil. This control voltage causes a solenoid action which closes the contacts of the device. These contacts may be connected in series with the power line going to a motor to cause it to turn on. The coil of the contactor is a relatively high resistance, causing it to draw a small current from the power line. This small current is capable of energizing the relay so that the closing of its contacts will apply power to a motor. Thus, a small current is used to control the larger current of the motor. The surface area of the contacts is usually fairly large to accommodate the current level drawn by the motor.

In this laboratory activity, you will use a magnetic contactor and push-button switch assembly to control the operation of a motor. Be sure to note the difference in magnitude of the control current through the contactor coil and the current drawn by the motor.

Equipment

Magnetic contactor—120 V.
Electric motor, any type—120 V.
Multifunction meter.
Variable ac power source.
Ac ammeter.
Push-button switch—normally open.
Push-button switch—normally closed.

Procedure

1. Construct the circuit shown in Fig. 5-21A.
2. Press the normally open push-button switch (START switch) to test the circuit for operation. The relay coil should energize, causing the motor to rotate.
3. Press the normally closed push-button switch (STOP switch). The motor should stop running.
4. Turn off the 120 volts ac power source and connect an ac ammeter in series with the ac line at point A.
5. Apply power to the circuit, press the START switch, and measure the starting and running currents of the motor.

   Starting current = ______ amperes ac.

   Running current = ______ amperes ac.
6. Turn off the circuit and disconnect the ac power source. Reconnect the ac ammeter in series with the relay coil at point B.
7. Apply power to the circuit, press the START switch, and measure the current through the coil.

   Coil current = ______ amperes ac.
8. With the ac line plugged into a variable 0–120-volt power source, perform the following steps.
9. Set the power source to zero volts and then turn it on. Keep the START button depressed while increasing the applied voltage. Measure and record the "pull-in" current and voltage of the contactor.

   Pull-in current = ______ amperes ac.

   Pull-in voltage = ______ volts ac.

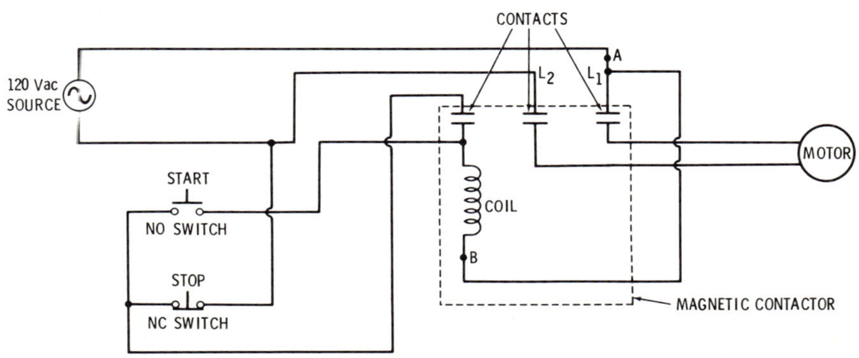

Fig. 5-21A. Magnetic contactor motor control circuit.

10. Gradually reduce the applied voltage until the contacts open. Record the "drop-out" current and voltage.

    Drop-out current = ______ amperes ac.

    Drop-out voltage = ______ volts ac.

11. This concludes the experiment.

Analysis

1. Why are magnetic contactors used to control motors?

____

____

____

2. Explain the operation of a magnetic contactor circuit to control a single-phase motor.

____

____

____

3. Draw a schematic diagram showing how a magnetic contactor could be used to control a three-phase motor.

4. What is meant by "pull-in" current?

____

____

____

5. What is meant by "drop-out" current?

____

____

____

5-22

Motor Control—Forward and Reverse Operation

Objective

Most types of electrical motors can be made to rotate in either direction by simple modifications of their connections. Ordinarily, motors used in industry require two magnetic contactors to accomplish forward and reverse operation. These contactors are used in conjunction with a set of three push-button switches—FORWARD, REVERSE, and STOP. When the FORWARD push-button switch is depressed, the forward contactor will be energized. It is deactivated when the STOP push-button switch is depressed. A similar procedure takes place during reverse operation.

In this laboratory activity, you will observe the forward and reverse operation of: (1) a dc shunt motor, and (2) a three-phase induction motor. The activity is designed for observation only; therefore, no measurements will be taken. Keep in mind the function of the magnetic contactor as a low-current control device for controlling the large currents drawn by motors.

Equipment

Dc shunt motor.
Three-phase induction motor.
Push-button switches (2)—normally open.
Push-button switch—normally closed.
Magnetic contactors (2).
Dc power source.
Three-phase ac power source.

Procedure

1. Construct the circuit shown in Fig. 5-22A for forward and reverse operation of a dc shunt motor.
2. Apply rated voltage to the dc shunt motor and check the circuit for operation. The FORWARD, REVERSE, and STOP switches should function. Have your instructor check the circuit for forward and reverse control.

   Instructor's Approval: ____________________

3. Disassemble the dc shunt motor control circuit and connect the circuit shown in Fig. 5-22B for forward and reverse operation of a three-phase induction motor.

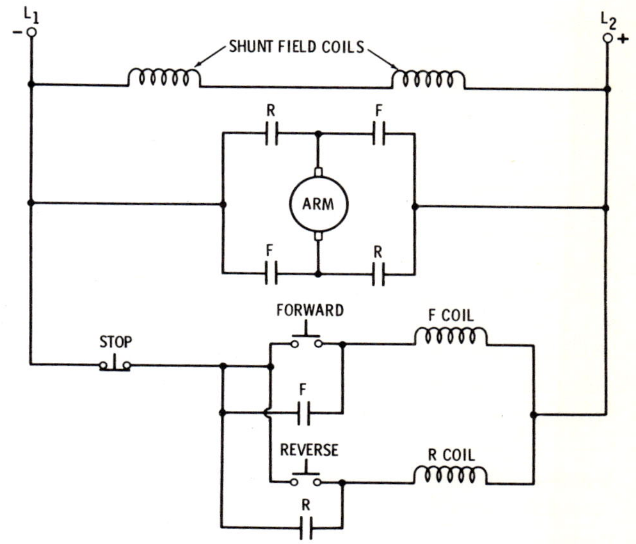

Fig. 5-22A. Control circuit for forward and reverse operation of a dc shunt motor.

4. Apply rated voltage to the three-phase motor and check the circuit for operation. The FORWARD, REVERSE, and STOP switches should function. Have your instructor check the circuit for forward and reverse operation.

   Instructor's Approval: ____________________

5. This concludes the experiment.

Analysis

1. How is the direction of rotation of a dc shunt motor changed?

   ____

   ____

   ____

2. How is the direction of rotation of a three-phase induction motor changed?

   ____

   ____

   ____

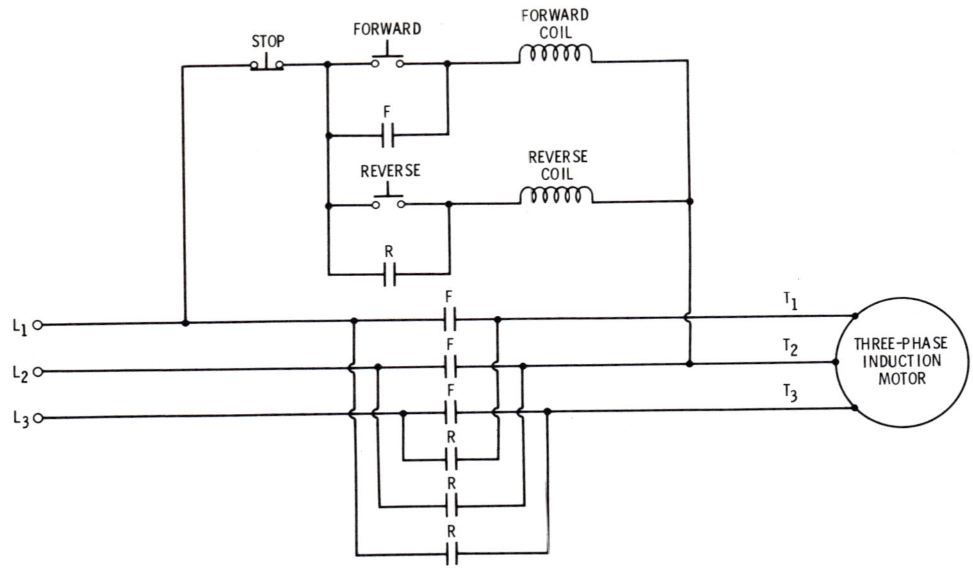

Fig. 5-22B. Control circuit for forward and reverse operation of a three-phase induction motor.

3. Explain the operation of the forward-reverse control of the dc shunt motor.

4. Explain the operation of the forward-reverse control of the three-phase induction motor.

5-23

Universal Motor Speed Control

Objective

An important advantage of a universal motor is the ease of speed control. Since this type of motor is similar to a series-wound dc motor, it has many of the same characteristics. The universal motor has a brush/commutator assembly, with the armature circuit connected in series with the field windings. By varying the voltage applied to universal motors, their speed may be varied from zero to maximum.

In this laboratory activity, you will observe the speed control characteristics of a universal motor. The circuit used for this purpose will be a gate-controlled triac. The triac is a semiconductor device whose conduction may be varied by a trigger voltage applied to its gate. A silicon controlled rectifier (SCR) could also be used as a speed control device for a universal motor. Speed control circuits like the one you will construct are used for many applications, such as electric drills, sewing machines, electric mixers, and several industrial applications.

Equipment

Universal motor.
Electronic multifunction meter.
Oscilloscope.
Resistors—68,000 Ω, 47,000 Ω, 4700 Ω.
Potentiometer—500,000 Ω.
Capacitors (2)—0.1 μF.
Diac—GE-X13.
Triac—GE-X12.

Procedure

1. Construct the universal motor speed control circuit shown in Fig. 5-23A.
2. Connect a properly synchronized oscilloscope across the triac (point A to point B) to observe the voltage across the device.
3. Adjust the 500,000-ohm speed control potentiometer through its range to check the circuit for operation. The motor speed should vary from zero to rated r/min.

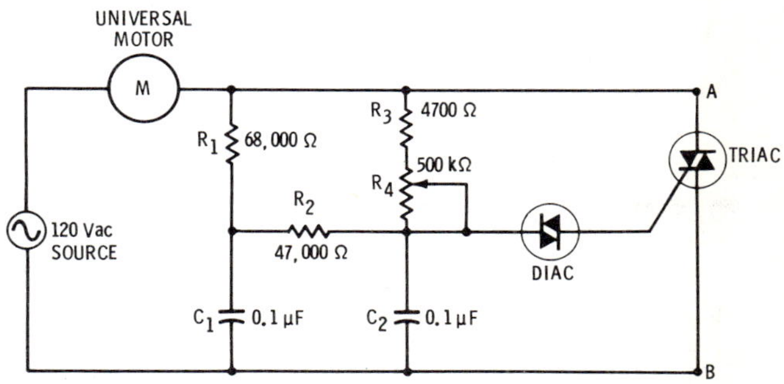

Fig. 5-23A. Speed control circuit for a universal motor.

4. Monitor the voltage across the triac as the potentiometer is adjusted. Observe the waveforms across the triac and the motor at the voltage levels indicated in Table 5-23A. Record these waveforms at each voltage setting.
5. This completes the oscilloscope observations.
6. Monitor the voltage across the potentiometer with a voltmeter. Record the maximum voltage as the potentiometer is varied, and observe its effect on the speed of the motor.

   Maximum voltage = ______ volts ac.

7. Monitor the voltage across capacitor C_2. Record the minimum and maximum voltage across C_2 as the potentiometer is varied.

   Minimum voltage = ______ volts ac.

   Maximum voltage = ______ volts ac.

8. This concludes the experiment.

Analysis

1. From the data of Table 5-23A, what is the relationship of the voltage across the motor and the voltage across the triac?

____

____

____

Table 5-23A. Speed Control Waveforms

Voltage Across Motor	Oscilloscope Waveforms	
	Across Motor	Across Triac
20 V ac		
40 V ac		
60 V ac		
80 V ac		
100 V ac		
120 V ac		

2. At what voltage across the motor does it begin to rotate?

3. How does the triac functionally control the speed of the universal motor?

4. From the data of Step 6, as potentiometer resistance increases, what happens to motor speed?

5. From the data of Step 7, as the voltage across capacitor C_2 increases, what happens to motor speed?

6. Why does this type of control allow the universal motor to produce more torque through its entire operational range than a rheostat control that reduces the applied ac voltage?

6-1

Current Measurement—Meter Shunts

Objective

The basic meter movement is a current-sensitive device. A 0–5-milliampere meter movement, for instance, requires 5 milliamperes of current to register full-scale deflection. A current of 2.5 milliamperes would cause one-half scale deflection. If values of current larger than 5 milliamperes were to be measured, this basic movement could not be used. However, the range of the meter can be extended to measure higher current values by adding a *shunt* resistor of a specified value in parallel with the movement.

In this laboratory activity, you will determine the resistance of the meter movement and then calculate shunt resistance values that can be used to extend the range of the meter movement to measure higher values of current. The activity is designed for a 0–100-microampere, 500-ohm, zero-centered meter movement; however, any meter movement can be used with values in the activity changed accordingly.

Equipment

Meter movement—0–100 μA, 500 Ω.
Multifunction meter.
Potentiometer—1000 Ω.
Variable dc power source.
Resistors—10,000 Ω, 2 W; 1200 Ω, 10 W.
Decade resistance box or selected resistance values.

Procedure

Section A. Meter Movement Resistance

1. Obtain a 0–100 μA, zero-centered meter movement.
2. Use the circuit shown in Fig. 6-1A to determine the resistance of the meter movement. This value cannot be measured directly with an ohmmeter.
3. Connect the meter movement in series with a 10,000-ohm resistor and dc power source as shown, with R_{SH} *out* of the circuit. Slowly increase the dc voltage until the meter reads full scale. Now, connect the potentiometer (R_{SH}) in parallel with the meter movement. Do not disturb the dc power supply setting.
4. Adjust the potentiometer until the meter movement reads half scale. Turn off the power, remove the potentiometer, and without disturbing its adjustment, measure its resistance. This resistance is equal to the meter movement resistance (R_M).

   R_M = ______ ohms.

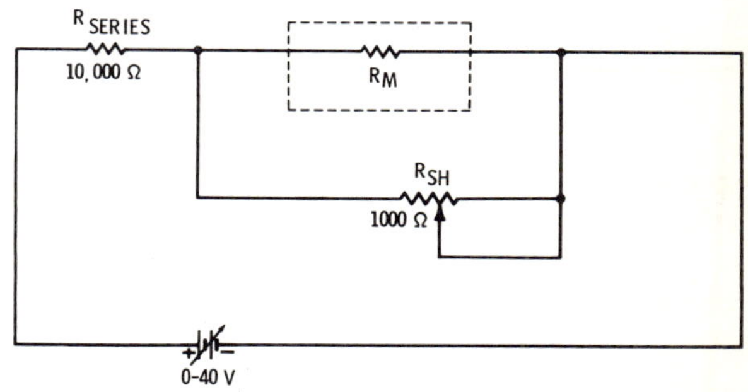

Fig. 6-1A. Circuit for determining the resistance of a meter movement.

5. Using the meter movement resistance (R_M) and the full-scale current (I_{FS}), write the meter movement specification:

   Movement specification = $I_{FS} \times R_M$ = ______ millivolts.

Section B. Meter Shunts for Current Measurement

1. Using the 100-μA meter movement and its resistance (R_M), calculate the values of shunt resistance (R_{SH})* needed to extend the range of the movement to measure:

   (a) 1.0 mA; R_{SH} = ______ ohms.

   (b) 10 mA; R_{SH} = ______ ohms.

   (c) 100 mA; R_{SH} = ______ ohms.
2. Use a decade resistance box, or necessary resistor combinations, to construct each of the shunts as shown in Fig. 6-1B. Use the values of R_{SH} calculated in Step 1, beginning with 1 mA.
3. Connect the circuits shown in Fig. 6-1B, and have your instructor check each R_{SH} value to verify that the meter movement will measure the values indicated. Be careful to set the voltage values correctly.

---

* $R_{SH} = \dfrac{I_M \times R_M}{I_{SH}}$

The shunts you have designed should indicate the values of $I = V/R$ for each of the three circuits.

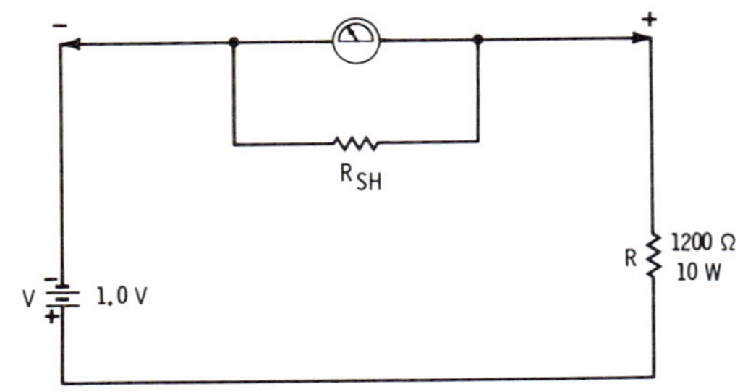

(a) For 1 mA; R_{SH} = ______ ohms.

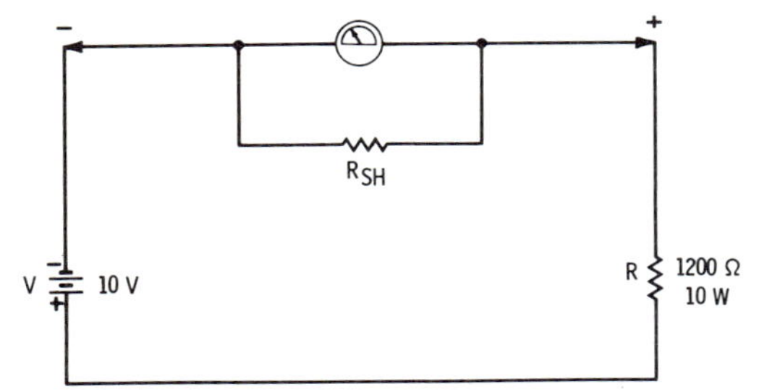

(b) For 10 mA; R_{SH} = ______ ohms.

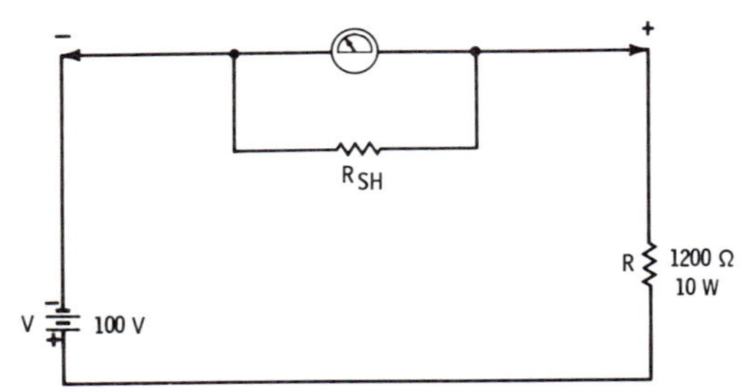

(c) For 100 mA; R_{SH} = ______ ohms.

Fig. 6-1B. Meter shunt test circuits.

Instructor's Approval: ______________________

(a) 1.0 mA: ______.

(b) 10 mA: ______.

(c) 100 mA: ______.

Analysis

1. Describe the construction of a hand-deflection meter movement such as the one used in this experiment.

____

____

____

2. If a 200-ohm, 1.0-mA meter movement is employed, what are the values of shunt resistance required to extend its range to measure the following currents? (Show your calculations.)

(a) 5 mA; R_{SH} = ______ Ω.

(b) 10 mA; R_{SH} = ______ Ω.

(c) 50 mA; R_{SH} = ______ Ω.

(d) 100 mA; R_{SH} = ______ Ω.

(e) 0.5 A; R_{SH} = ______ Ω.

(f) 1.0 A; R_{SH} = ______ Ω.

3. Draw a diagram of a meter-movement circuit that could be used to measure alternating-current values.

6-2

DC Voltage Measurement—Meter Multipliers

Objective

In this activity, you will use a basic meter movement to measure dc voltage. In the previous activity, the meter movement was used to measure current. In a similar manner, the basic meter movement can be used to measure voltage values. When a voltage is applied to the meter movement, a certain amount of current flows through the movement. For example, if the 0–5-mA movement discussed in the previous activity had an internal resistance of 100 ohms, and 0.5 volt were applied to it, full-scale deflection would result. This can be verified by using Ohm's law. The voltmeter movement can also be extended to measure higher values. In this case, a *multiplier* resistor is placed in series with the movement.

Equipment

Meter movement—0–100 μA, zero-centered, 500 Ω.
Dc power source.
Decade resistance box or assorted resistors.
Multifunction meter.

Procedure

1. The circuit shown in Fig. 6-2A illustrates how a meter movement is used to measure voltages.

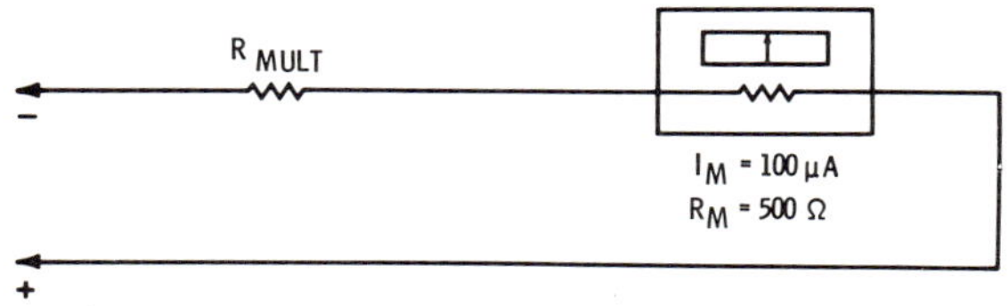

Fig. 6-2A. Basic meter movement circuit for measuring voltages.

2. Using the 0–100-μA, 500-ohm meter movement, calculate the value of multiplier resistance needed to extend the meter range to measure the following voltages:

(a) 2 volts; R_{MULT} = ______ ohms.

(b) 5 volts; R_{MULT} = ______ ohms.

(c) 10 volts; R_{MULT} = ______ ohms.

(d) 25 volts; R_{MULT} = ______ ohms.

3. Have your instructor check these values as you verify them by applying the indicated voltages to the meter circuits of Fig. 6-2B with your calculated values of R_{MULT}. *Be careful.*

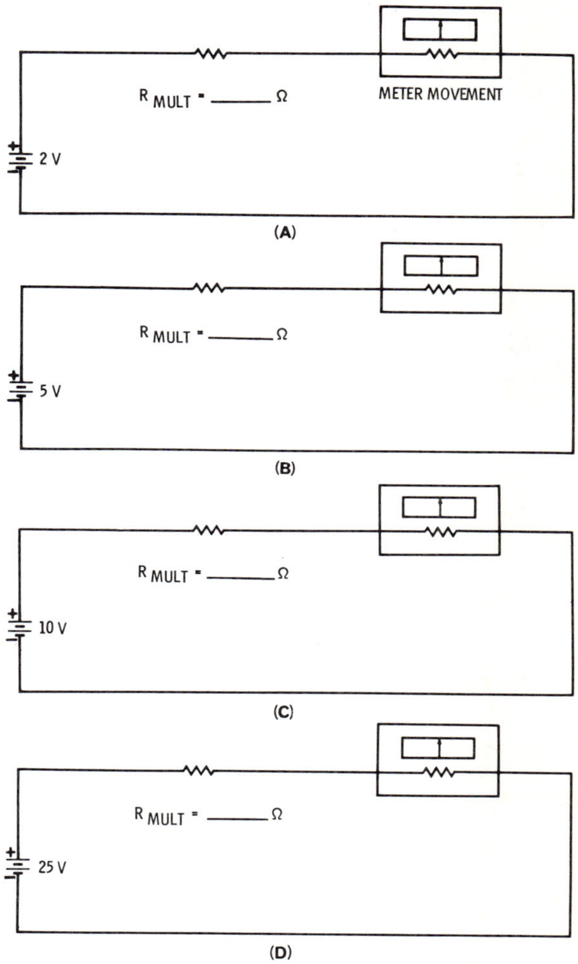

Fig. 6-2B. Meter multiplier test circuits.

Instructor's Approval: ____________________

(a) 2 V: ______.

(b) 5 V: ______.

(c) 10 V: ______.

(d) 25 V: ______.

4. This concludes the experiment.

Analysis

1. Calculate the percentage difference between the measured values of voltage in Step 3 and the actual values.

2. Compute the "ohms-per-volt" rating of the meter movement you used in this experiment.

3. What is meant by meter "loading" of a circuit?

____

____

____

4. Calculate the values of multiplier resistance needed to extend the range of a 1.0-mA, 100-ohm meter movement to measure the following voltages:

(a) 1.0 volt; R_{MULT} = ______ Ω.

(b) 5 volts; R_{MULT} = ______ Ω.

(c) 15 volts; R_{MULT} = ______ Ω.

(d) 150 volts; R_{MULT} = ______ Ω.

Wheatstone Bridge Measurements

Objective

Bridge circuits are widely used for the measurement of resistance, capacitance, inductance, impedance, frequency, and other circuit parameters. A bridge circuit provides a comparison of an *unknown component* in terms of an accurate *known standard component.*

A most common type of bridge circuit is the *Wheatstone bridge,* which will be investigated in this activity. This circuit has four resistive arms, a source of voltage, and a *null detector* (usually a galvanometer or sensitive current meter). The current through the meter depends on the potential difference between the points on the bridge. The bridge is balanced when there is no current flow through the meter and the potential difference is zero volts.

Equipment

Resistors—1000 Ω, 220 Ω, 100 Ω.
Potentiometers—10,000 Ω, 1000 Ω.
Galvanometer.
Variable dc power source.
Multifunction meter.
Spst switches (2).

Procedure

1. Construct the Wheatstone bridge circuit shown in Fig. 6-3A.
2. Turn on the power supply and apply 10 volts dc to the bridge.
3. Close switch S_2 to monitor the current flowing through the galvanometer.
4. Observe and record the initial meter reading:

   ______ μA dc.
5. Adjust potentiometer R_S until a null or zero current reading is indicated on the meter.
6. Measure the following voltages with the multifunction meter:

   (a) Points A to B: ______ volts dc.

   (b) Points A to C: ______ volts dc.

   (c) Points B to C: ______ volts dc.

   (d) Points A to D: ______ volts dc.

   (e) Points B to D: ______ volts dc.
7. Turn off the power source and check the resistance setting of R_S.

   R_S = ______ ohms (measured value) at the null condition.

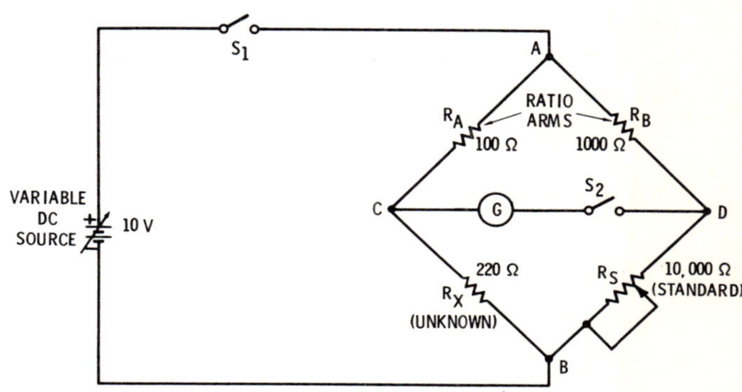

Fig. 6-3A. Wheatstone bridge circuit.

8. Replace R_X with a 0–1000-ohm potentiometer. Apply power to the bridge. Record the range of the galvanometer readings as R_S is adjusted:

   ______ to ______ μA dc.
9. Adjust R_X to midrange.
10. Adjust R_S until a null reading is indicated on the meter.
11. Measure the following voltages:

    (a) Points A to B: ______ volts dc.

    (b) Points A to C: ______ volts dc.

    (c) Points B to C: ______ volts dc.

    (d) Points A to D: ______ volts dc.

    (e) Points B to D: ______ volts dc.
12. Turn off the power supply and check the resistance settings of R_X and R_S.

    R_X = ______ ohms.

    R_S = ______ ohms.
13. This concludes the experiment.

Analysis

1. From Step 5, what is the relationship between the two branches of the bridge network?

2. From Step 6, how does the voltage drop across R_S compare to the voltage drop across R_X? How does the drop across R_A compare to the drop across R_B?

3. For this circuit, the relationship $R_X = (R_A/R_B) \times R_S$ is used. Use the values of this activity to verify this relationship.

4. Compare the actual and calculated values of R_X from the first part of the activity.

5. Calculate the value of R_X for the last part of the activity. Compare this value to the measured value.

6. Briefly describe the operation of the Wheatstone bridge.

6-4

Oscilloscope Measurements

Objective

The oscilloscope is one of the most important laboratory instruments in use today. It is essential for an industrial technician to be able to use the oscilloscope competently. Oscilloscopes can be used to measure voltages, frequency, phase relationships, or to monitor waveform displays of various types.

In this activity, you will use the oscilloscope to measure the amplitude of an ac sine-wave voltage. The amplitude of an ac sine wave can be measured in three different ways: (1) peak-to-peak (p-p), (2) peak, and (3) root-mean-square (rms) or effective value. Peak-to-peak, as the term implies, is the measurement from the peak of the positive half of the waveform to the peak of the negative half, while peak is simply the measurement of either the positive or negative half. Rms (effective) voltage is the measurement of the amount of ac voltage that will do the same amount of work as a comparable dc voltage. The power-line voltage in the home, for example, is stated as an rms value. Most ac meters can only accurately measure rms sine-wave voltages, but the oscilloscope can measure all voltages regardless of waveform type. This is a primary advantage of the oscilloscope.

Equipment

Oscilloscope.
Ac power source.
Resistors—270 Ω, 470 Ω, 1000 Ω.
Multifunction meter.

Procedure

1. Turn on the power supply and adjust the voltage to 10 volts ac. Calculate the peak and p-p values of this voltage.

   Peak value = ______ volts ac.

   P-p value = ______ volts ac.
2. Construct the circuit shown in Fig. 6-4A and connect it to the 10-volt ac source. Measure the voltage across the following points with the meter. Calculate the peak and p-p values.

   (a) Points A to B: ______ volts rms; ______ volts peak; ______ volts p-p.

   (b) Points B to C: ______ volts rms; ______ volts peak; ______ volts p-p.

   (c) Points C to D: ______ volts rms; ______ volts peak; ______ volts p-p.
3. Prepare the oscilloscope for operation by adjusting the appropriate controls. Connect the ground lead to point D and the vertical-input lead to point A. Properly center the waveform and adjust the horizontal frequency and sync controls to produce two sine waves. Adjust the vertical controls so that each square or horizontal line of the graticule on the scope screen equals a definite p-p voltage. As previously determined mathematically, the p-p voltage of the 10-volt source is approximately 28.2 volts. Calibrate the scope graticule so the sine wave has a specific number of squares or lines of amplitude. Once the scope has been calibrated, do not adjust the vertical controls again. (Note: Some scopes have internal calibration and the preceding is not required.)
4. With the scope calibrated, measure the p-p voltage across the following points and record the values:

   (a) Points A to B = ______ volts p-p.

   (b) Points B to C = ______ volts p-p.

   (c) Points C to D = ______ volts p-p.
5. This concludes the activity.

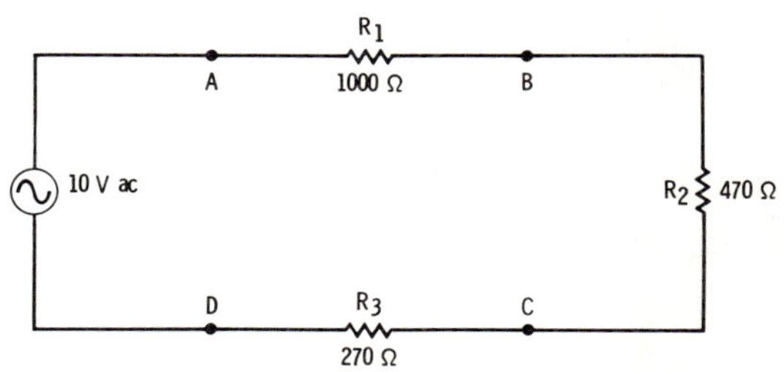

Fig. 6-4A. Test circuit for ac voltage measurements.

Analysis

1. In order to measure p-p voltage, the scope must first be calibrated, then it can be used to make any p-p measurements. How does changing the vertical attenuation or volts/division control affect the calibration?

2. Discuss the basic operation of the oscilloscope.

3. How can an oscilloscope be used to measure frequency?

4. Explain the procedure for calibrating an oscilloscope to measure p-p ac voltage.

6-5

Chart Recorder Familiarization

Objective

Chart recorders are used to monitor many types of electrical and physical quantities over a period of time. The advantage of a chart recorder is that a visible record of the monitored quantity may be kept for future use. There are many different types of chart recorders in use today.

In this laboratory activity, you will be able to study the construction and operation of a chart recorder. The recorder should be connected to some operational system so that you can observe its operation.

Equipment

Chart recorder (any type available in the lab).

Procedure

1. Obtain a chart recording instrument and operating manual (if available) from your instructor. Record the manufacturer and model number.

   Manufacturer ______________________.

   Model Number ______________________.

2. What quantity or quantities does this instrument measure?

   ____

   ____

   ____

3. What is the range or scale of this instrument?

   ____

   ____

   ____

4. What method is used by this instrument to make a permanent record of the quantity measured?

   ____

   ____

   ____

5. What type of drive mechanism is used on this instrument?

   ____

6. What type of chart is used by this instrument?

   ____

7. Describe any special features this instrument has:

   ____

   ____

   ____

8. Prepare the chart recorder for operation. Connect the recorder to some circuit or system in which you can observe its operation over its entire range of measurement.

9. Have your instructor check the instrument for operation.

Instructor's Approval: ______________________

10. This concludes the laboratory activity.

Analysis

1. What are some specific applications where chart recorders might be used?

   ____

   ____

   ____

2. What are some types of drive mechanisms used for chart recorders?

   ____

   ____

   ____

3. What are some methods used by chart recorders for recording the measured quantity.

   ____

   ____

   ____

6-6

Single-Phase Power Measurement

Objective

In direct-current circuits, power (watts) is always equal to the product of voltage and current. This relationship is also true for resistive ac circuits. However, when reactive circuits that contain either inductance or capacitance are encountered, the power converted in the circuit is no longer equal to the product of voltage and current. It is necessary to use a wattmeter to measure the *true power* (W) of these types of ac circuits. The product of voltage and current is referred to as the *apparent power* (VA) of a circuit. The power delivered from the source, but not converted to another form of energy by the load, is called *reactive power* (var). These three components form the power triangle relationship.

In this laboratory activity, you will use a single-phase wattmeter to measure the true power of a motor.

Equipment

Single-phase wattmeter (2000 W or higher).
Single-phase ac motor.
Multifunction meter.
Ac current meter.
120-V ac single-phase power source.

You will be able to measure the voltage, current, and power associated with the operation of a single-phase ac motor

Procedure

1. Obtain a single-phase ac motor of any type, and a single-phase wattmeter.

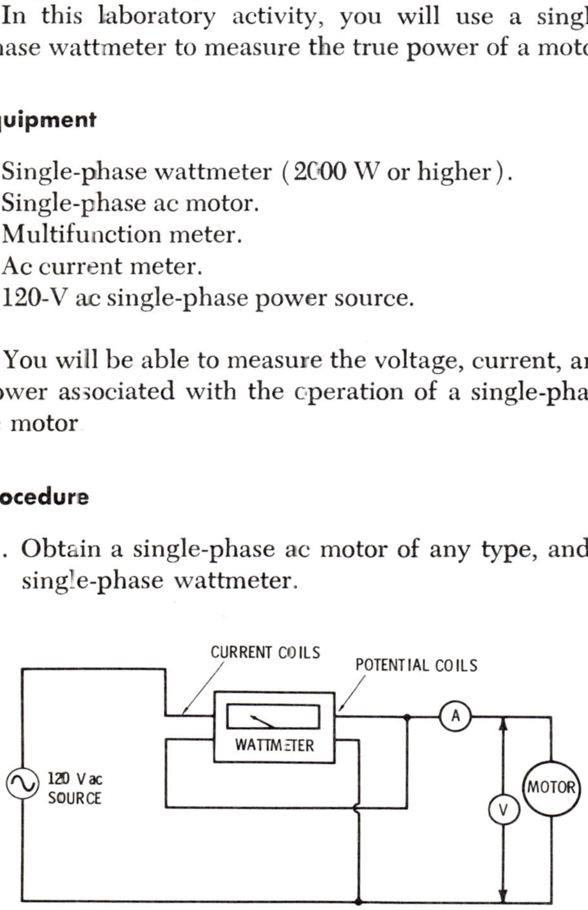

Fig. 6-6A. Test circuit for single-phase power measurements.

2. Check the nameplate data on the motor. Multiply the rated voltage of the motor by its rated current. Record the values.

   Rated voltage = ______ volts ac.

   Rated current = ______ amperes ac.
3. Multiply the voltage and current values to obtain a full-load apparent-power rating.

   Apparent power = ______ VA.
4. Make sure that the wattmeter will measure this amount of power.
5. Connect the wattmeter to the motor as shown in Fig. 6-6A.
6. Measure and record the ac voltage (V) applied to the motor.

   Voltage = ______ volts ac.
7. Apply power to the motor. Observe the reading on the wattmeter and current meter. Record the maximum power and current during the starting period.

   Starting power = ______ watts.

   Starting current = ______ amperes ac.
8. Measure and record the power and current while the motor is running.

   Power = ______ watts.

   Current = ______ amperes ac.
9. Apply a load to the shaft of the motor while observing the meters. Record the maximum power and current.

   Maximum power = ______ watts.

   Maximum current = ______ amperes ac.
10. This concludes the experiment.

Analysis

1. How is a single-phase wattmeter constructed?

______

______

______

2. From the data of Step 7, calculate:
   (a) Apparent power = volts × amperes = ________ VA.

   (b) Power factor = $\frac{\text{true power}}{\text{apparent power}}$ = ________.

   (c) Reactive power = $\sqrt{VA^2 - W^2}$ = ________ var.

3. From the data of Step 8, calculate:

   (a) Apparent power = ________ VA.

   (b) Power factor = ________.

   (c) Reactive power = ________ var.

4. From the data of Step 9, calculate:

   (a) Apparent power = ________ VA.

   (b) Power factor = ________.

   (c) Reactive power = ________ var.

5. As the load is increased on the single-phase ac motor, what happens to:
   (a) Apparent power?

   ____

   (b) True power?

   ____

   (c) Power factor?

   ____

   (d) Reactive power?

   ____

6-7

Three-Phase Power Measurement

Objective

Two single-phase wattmeters can be used to measure the power converted by a three-phase load. When the meters are properly connected, the algebraic sum of the two readings will equal the total three-phase power converted by the load. However, if the power factor of the load device is less than 50%, one wattmeter will deflect in the wrong direction. In this case, the current-coil connections on the meter should be reversed so that the meter will deflect properly. Then, the lower meter reading will be subtracted from the higher one.

In this laboratory activity, you will use two single-phase wattmeters to measure the power converted by a three-phase motor.

Equipment

Single-phase wattmeters (2).
Three-phase ac motor.
Multifunction meter.
Ac current meter.
Three-phase power source.

Procedure

1. Obtain a 208-volt or a 240-volt three-phase ac motor and connect two single-phase wattmeters to it as shown in Fig. 6-7A.
2. Measure and record the ac voltage applied to the motor.

   Voltage = _______ volts ac.

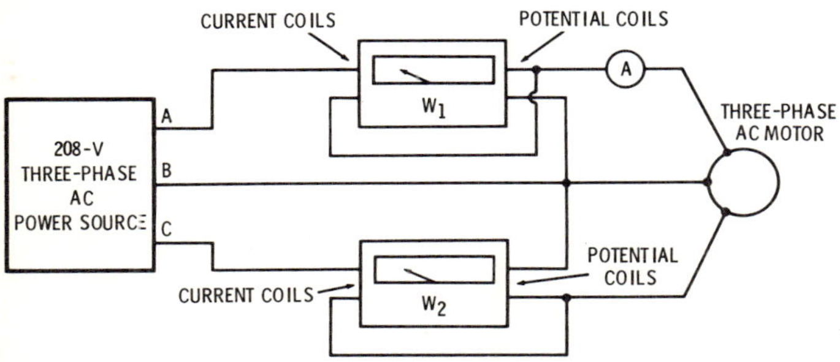

Fig. 6-7A. Test circuit for three-phase power measurements.

3. Apply power to the motor. Observe the wattmeter and ammeter readings. Record the maximum value of each when the motor starts.

   Starting power (wattmeter 1) = _______ watts.

   Starting power (wattmeter 2) = _______ watts.

   Starting current = _______ amperes ac.
4. Measure and record the same values while the motor is running.

   Power (wattmeter 1) = _______ watts.

   Power (wattmeter 2) = _______ watts.

   Current = _______ amperes ac.
5. Apply a load to the motor shaft while observing the meters. Record the maximum power and current.

   Maximum power (wattmeter 1) = _______ watts.

   Maximum power (wattmeter 2) = _______ watts.

   Maximum current = _______ amperes ac.
6. This concludes the experiment.

Analysis

1. What are some ways in which three-phase power can be measured?

   ____

   ____

   ____
2. From the data of Step 3, calculate:

   (a) True power = $W_1 + W_2$ = _______ watts.
   (b) Apparent power = $V \times A \times 1.73$ =

   _______ VA.

   (c) Power factor = $\frac{\text{true power}}{\text{apparent power}}$ = _______.

   (d) Reactive power = $\sqrt{VA^2 - W^2}$ = _______ var.

3. As load is increased on a three-phase motor, what happens to:
   (a) Apparent power?

   ____

   (b) True power?

   ____

   (c) Power factor?

   ____

   (d) Reactive power?

   ____

4. From the data of Step 4, calculate:
   (a) True power = ______ watts.
   (b) Apparent power = ______ VA.
   (c) Power factor = ______.
   (d) Reactive power = ______ var.

5. From the data of Step 5, calculate:
   (a) True power = ______ watts.
   (b) Apparent power = ______ VA.
   (c) Power factor = ______.
   (d) Reactive power = ______ var.

6-8

Watt-Hour Meter Familiarization

Objective

Watt-hour meters are used to monitor the amount of electrical power used over a period of time. Any industrial plant or residence that uses electrical power has a watt-hour meter to measure the power used at that particular location.

In this laboratory activity, you will be able to observe the operation of a watt-hour meter. You may connect a load to the meter (such as a motor) to cause it to function. If a meter is not available in the lab, you can observe an operational watt-hour meter either near the laboratory or at home.

Equipment

Single-phase watt-hour meter.
Single-phase ac motor or other load device.

Procedure

1. Connect the watt-hour meter to the load as shown in Fig. 6-8A.
2. Apply power to the load and observe the operation of the watt-hour meter. Note the speed of rotation of the aluminum disk.

   Speed = ______ r/min (approximately).
3. Increase the load while observing the speed of the aluminum disk.

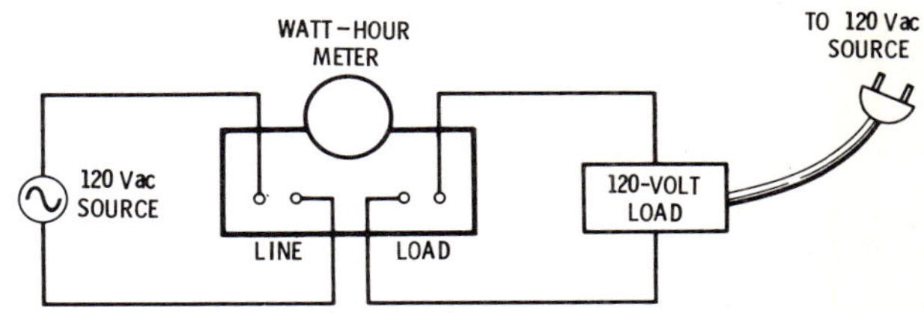

Fig. 6-8A. Circuit for observing the operation of a watt-hour meter.

4. Have your instructor check your meter for operation.

Instructor's Approval: ____________________

5. This concludes the experiment.

Analysis

1. Discuss the operational principle of a watt-hour meter.

   ____
   ____
   ____
2. In Step 3, what happened to the speed of the aluminum disk? Why?

   ____
   ____
   ____

6-9

Telemetering Systems

Objective

In this activity, you will construct a simple circuit that illustrates a telemetering system. Telemetering takes place when a change in one electrical or physical variable is transmitted by some means over a relatively long distance from where the change occurs. The amount of change is noted on the receiving end of the system by observation of some type of recording or monitoring device.

This activity uses two 555 IC timers. They are used to sense a change in temperature of a thermistor, which causes a frequency change in the circuit. This frequency change can be noted on an am radio receiver located a long distance away.

The 555 IC can produce time delays from a few microseconds to hours. It can run as an oscillator at frequencies as low as one pulse per hour or as high as one megahertz. The internal structure of a 555 is composed of many transistors and other components, arranged to make up the following circuits: one bistable multivibrator (flip-flop), two comparators, and an output stage. The oscillation capability of the 555 is used in this activity. Its frequency is changed by an external RC network in which the resistance is a temperature sensitive thermistor. Thus, changes in temperature can be transmitted over a long distance to an am radio receiver and noted as changes in frequency.

Equipment

Integrated circuits—555 IC timers (2).
Oscilloscope.
Dc power source.
Capacitors—24 pF, 0.01 μF, 1.5 μF.
Resistors—650 Ω, 1000 Ω (2).
Thermistor—1000 Ω (KA31L1 or equivalent).
Am radio receiver.
Wire—5 feet (for antenna).

Procedure

1. Construct the simple telemetering circuit shown in Fig. 6-9A.
2. Using the oscilloscope, record the waveform produced at the antenna of the circuit.

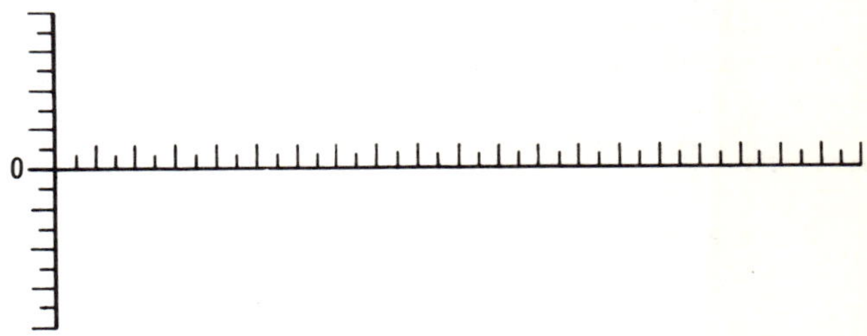

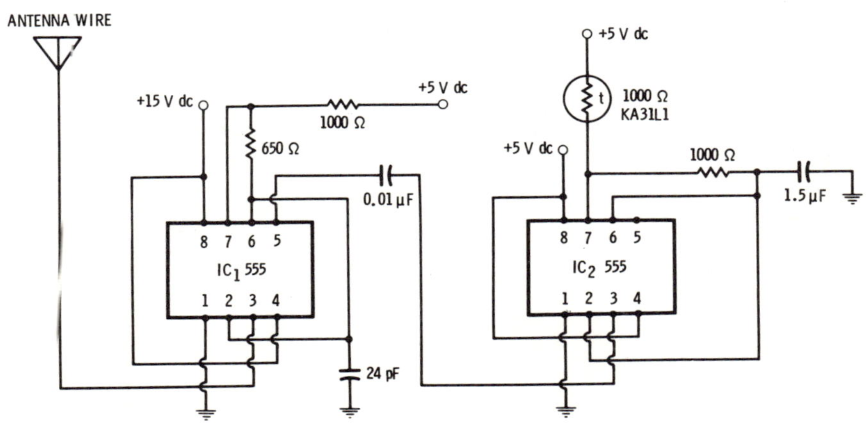

Fig. 6-9A. Telemetering circuit.

3. Using the oscilloscope, determine the frequency of the waveform produced at the antenna.

   Frequency = ______ Hz.
4. Record the waveform produced at the output of IC_2 (pin 3).

5. Determine the frequency of the waveform produced at the output of IC_2.

   Frequency = ______ Hz.

   NOTE: The frequency as measured at the antenna must be between 550 kHz and 1650 kHz.
6. Obtain an am radio receiver and place it about 5 to 10 feet from the antenna. Adjust the tuner of the radio until the tone produced by the circuit becomes audible.

   NOTE: The tuner of the radio should be adjusted to the frequency measured in Step 3.
7. Change the temperature of the thermistor connected to IC_2 by grasping it with your fingers. Observe the result of the temperature change.
8. This concludes the experiment.

Analysis

1. Write a brief analysis of this telemetry system.
2. What are some possible applications of the principle of telemetering illustrated in this activity as an instrumentation method?
3. How do the frequencies of the waveforms of Steps 2 and 4 compare? Why?
4. In Step 7, what effect did an increase in temperature of the thermistor cause?
5. What are some types of telemetering systems used in industry?

6-10

Transducers—Strain Gauges

Objective

The strain gauge is a commonly used resistive transducer. It is used in the measurement of displacement, torque, weight, pressure, and other force parameters. In this activity, a strain gauge used in a bridge network will be observed. This method could be used for most strain gauge measurements.

A strain gauge has a fine-wire element looped back and forth on a mounting plate, which is usually cemented to the member undergoing stress. The extra length due to the looping increases the effect of any stress applied lengthwise. A tensile stress stretches the wire, increasing its length and, therefore, its resistance.

The measure of the sensitivity of a material to strain is called the *gauge factor* (GF), and is the ratio of the change in resistance (R/R) to the change in length (L/L) or:

$$GF = \frac{\Delta R/R}{\Delta L/L}$$

The initial resistance (R) is typically around 120 ohms, and the gauge factor may run from −12 to +6. A gauge factor of 2 is common for most wire strain gauges.

The more recent semiconductor strain gauges have greater sensitivity and can be used exactly like a metallic strain gauge. Semiconductor strain gauges have a gauge factor roughly 50 times greater (around 100) than the metallic types. This large gauge factor is accompanied by a thermal rate of change of resistance also approximately 50 times higher than in older gauges. Thus, the semiconductor strain gauge is as stable as the metallic type, but has a much higher output. Simpler temperature compensation methods can be applied to the semiconductor strain gauge, making it more useful and extending its range to the measurement of small values of microstrain (microinches/inch of length), which could not be measured with earlier gauges.

Equipment

Resistors—1000 Ω, 100,000 Ω.
Strain gauge.
Helipot (calibrated)—20,000 Ω.
Dc power source.
Galvanometer.
Multifunction meter.

Procedure

1. Construct the strain gauge balance detector circuit shown in Fig. 6-10A.

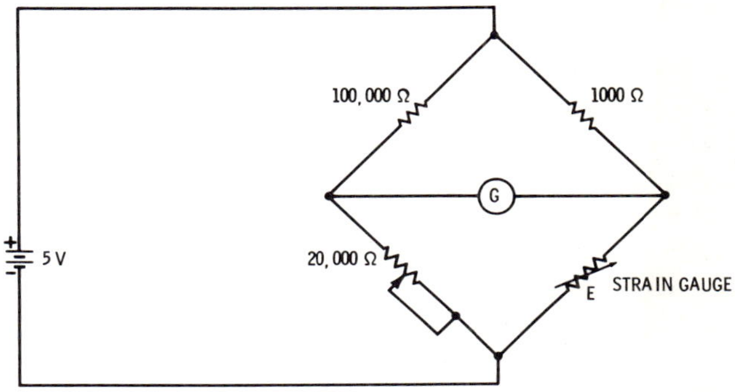

Fig. 6-10A. Strain gauge balance detector test circuit.

2. Measure the resistance of the strain gauge with the meter.

   Resistance = ______ ohms.

3. While your lab partner bends the metal mounting plate, monitor the null detector and observe the effect on the meter reading.
4. This concludes the activity.

Analysis

1. Write an analysis of the operation of the strain gauge. Include in your analysis your observation of the bridge network as a measuring method for the strain gauge.

6-11

Transducers—Thermocouple Characteristics

Objective

In this laboratory activity, you will observe the characteristics of a type J thermocouple. The thermocouple is an often-used, heat-sensitive transducer. An increase in temperature causes the millivolt output of the thermocouple to increase proportionally. This relationship is expressed in the temperature-millivolt characteristic curves of Appendix 3.

Equipment

Thermocouples—type J (2).
Meter (capable of measuring millivolts).
Appendix 3—Thermocouple Temperature-Millivolt Curves.
Heating coil and socket—660 W.
Soldering iron.

Procedure

1. Acquire a type J thermocouple and a voltmeter. Construct the circuit shown in Fig. 6-11A.

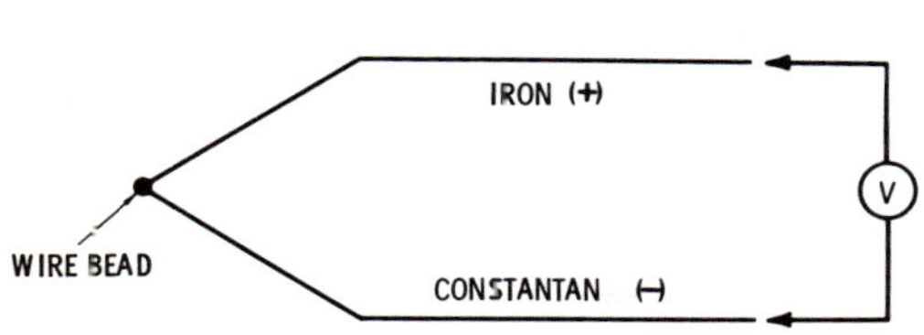

Fig. 6-11A. Thermocouple characteristics test circuit.

2. Obtain a 660-watt heating coil and connect it to 120 volts ac. Allow it to warm up for about three minutes. (Note: This cone is *very hot*. Extreme caution should be exercised when working with this device.)
3. Place the wire-bead junction of the thermocouple within 1/4 inch of the heating coil. Complete Table 6-11A using the temperature-millivolt curves of Appendix 3.

Table 6-11A. Temperature-Millivolt Characteristics

Millivolts (mV)	°F	°C	K	°R

4. Disconnect the heat cone and allow it to cool.
5. Obtain a soldering iron and connect it to 120 volts ac. Allow it to warm up for about five minutes.
6. Place the wire-bead junction of the thermocouple within 1/4 inch of the *hot* soldering iron tip and complete Table 6-11B using the temperature-millivolt curves.

Table 6-11B. Temperature-Millivolt Characteristics

Millivolts (mV)	°F	°C	K	°R

7. Again, use the 660-watt heating coil. Connect it to 120 volts ac and allow it to warm up for about three minutes.
8. Place the wire bead junction of the thermocouple within 1/4 inch of the coil. Measure and record the time in seconds required for the thermocouple to reach its maximum output.

   Time = ______ seconds.
9. Construct the circuit shown in Fig. 6-11B.

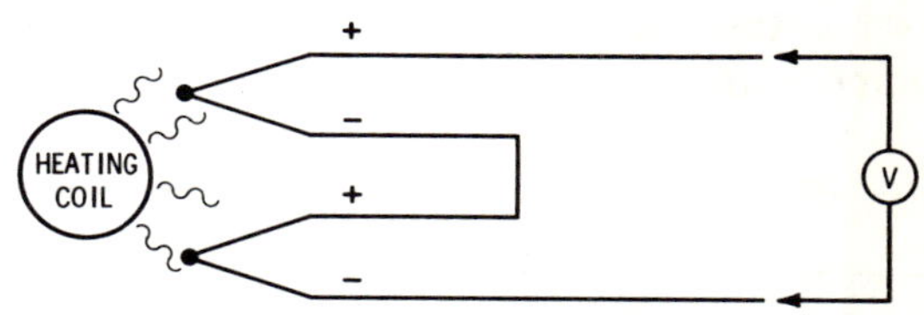

Fig. 6-11B. Thermocouples in series.

10. Allow the heating coil to warm to maximum temperature and place the wire bead junctions of the two thermocouples within 1/4 inch of the coil. Complete Table 6-11C.

Table 6-11C. Temperature-Millivolt Series Characteristics

Millivolts (mV)	°F	°C	K	°R

11. Construct the circuit shown in Fig. 6-11C.

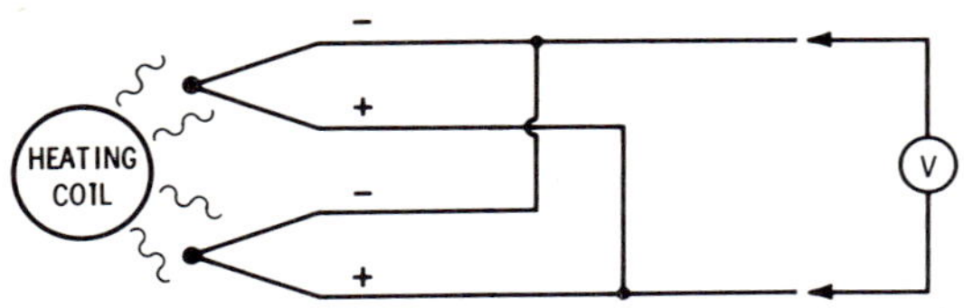

Fig. 6-11C. Thermocouples in parallel.

12. Repeat Step 10 and complete Table 6-11D.

Table 6-11D. Temperature-Millivolt Parallel Characteristics

Millivolts (mV)	°F	°C	K	°R

Analysis

1. In the following space, design and draw a circuit, using one or more thermocouples, that would measure the temperature of a 60-watt light bulb.

2. Discuss the characteristics and operation of the thermocouple.

3. Using the data of Step 8, determine the time-constant code of this thermocouple.

4. How did the data collected in Step 10 compare with the data of Step 3? Why?

5. How did the data of Step 12 compare with that of Step 10? Why?

6. What are some possible industrial applications of a series connection?

7. What are some possible industrial applications of a parallel connection?

6-12

Transducers—Thermocouple Applications

Objective

In this activity, you will construct a circuit that uses a thermocouple as a transducer. The thermocouple is connected to the gate of a field-effect transistor (FET). Any change in temperature of the thermocouple will cause a corresponding change in FET conduction characteristics.

Equipment

Thermocouple—type J.
FET—GE-FET-1.
Dc power sources (2).
Resistors—680 Ω, 6800 Ω.
Multifunction meter.
Heating coil—660 W.
Relay.
Transistors—2N3053 (2).

Procedure

1. Construct the thermocouple transducer circuit shown in Fig. 6-12A.

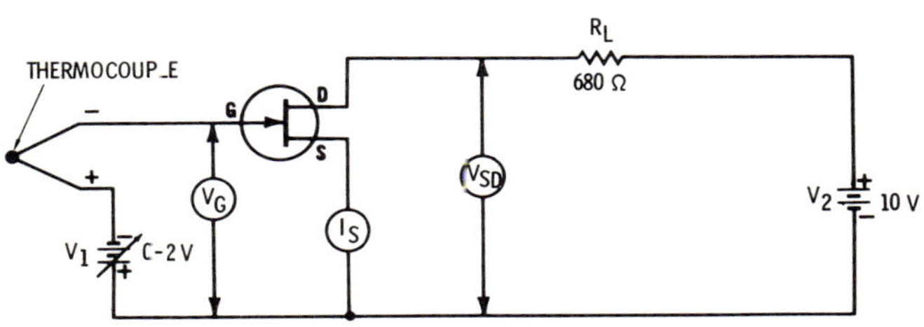

Fig. 6-12A. Thermocouple transducer test circuit.

2. Allow the measurement junction of the thermocouple to remain at room temperature and complete Table 6-12A. Adjust V_1 to obtain the range of gate voltage.
3. Acquire the 660-watt heating coil and connect it to 120 volts ac. Allow it to warm up for about three minutes.
4. Adjust the gate voltage of the FET to 1.0 volt dc.
5. Place the metal bead junction of the thermocouple inside the heat cone and record the following:

Table 6-12A. Circuit Characteristics at Room Temperature

Gate Voltage (V_G)	Drain Voltage (V_{SD})	Source Current (I_S)	V_{R_L}
0.2			
0.4			
0.6			
0.8			
1.0			
1.5			
2.0			

V_{SD} = ________.

I_S = ________.

V_{R_L} = ________.

6. Disconnect the heat cone and allow the thermocouple to return to room temperature.
7. Construct the circuit illustrated in Fig. 6-12B.
8. Allowing the measurement junction to remain at room temperature, adjust the gate voltage of the FET until the relay activates.
9. Adjust the gate voltage until the relay is ready to drop out.
10. Connect the 660-watt heating coil to 120 volts ac and allow it to warm up for three minutes.
11. Place the metal bead junction into the heating coil. Observe the reaction of the relay.
12. Remove the power to the heating coil and allow both the coil and the metal bead junction to cool to room temperature. Observe the reaction of the relay.

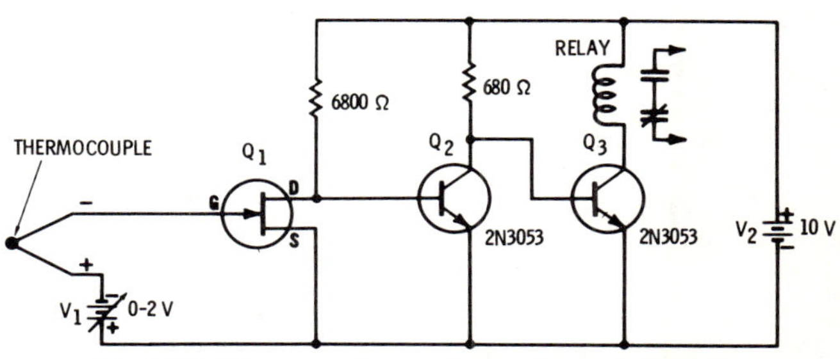

Fig. 6-12B. Thermocouple transducer relay control circuit.

Analysis

1. How does the thermocouple affect the biasing of the FET?

2. From the data of Table 6-12A, as gate voltage increases, what happens to:
   (a) Drain voltage (V_{SD})?

   (b) Source current (I_S)?

   (c) Load voltage (V_{R_L})?

3. In Step 5, what happens to V_{SD}, I_S, and V_{R_L} when the thermocouple is heated?

4. How do you account for the differences noted in Step 5?

5. In Step 11, what occurred when the thermocouple was heated?

6. In Step 12, what occurred when the thermocouple was cooled?

7. What is the purpose of Q_2 and Q_3 in the circuit of Fig. 6-12B?

6-13

Transducers—Thermistor Characteristics

Objective

In this laboratory activity, you will observe the characteristics of one type of thermistor. Thermistors are commonly used as temperature sensitive transducers. Their most important property is that their resistance decreases with increases in temperature.

Equipment

Thermistor–JA41J1.
Multifunction meter.
Dc power source.
Resistor—1000 Ω.
Lamp--60 W.

Procedure

1. Construct the circuit shown in Fig. 6-13A.

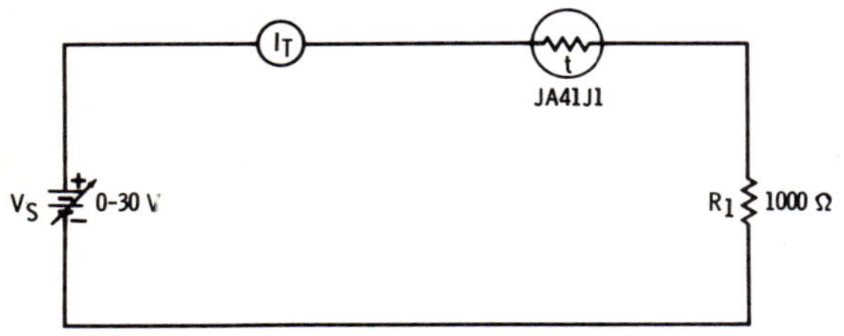

Fig. 6-13A. Thermistor characteristics test circuit.

2. Complete Table 6-13A relative to the circuit of Fig. 6-13A with the thermistor at room temperature: (Note: Allow about two minutes between source voltage changes.)
3. Repeat Step 2 using a 60-watt lamp placed about one inch from the thermistor. This will increase the temperature and change the characteristics of the thermistor. Again, allow about two minutes between voltage changes. Record your data in Table 6-13B.
4. Allow the thermistor to return to room temperature.
5. Apply 0.5 volt dc to the circuit and record the current caused by the circuit with the thermistor at room temperature.

   Current = ______ milliamperes.
6. Grasp the thermistor with your hand for three minutes and observe what happens to the current in the circuit.
7. This concludes the activity.

Analysis

1. Draw a current versus voltage graph to illustrate the characteristics of the thermistor from the data of Tables 6-13A and B.

2. Briefly discuss the characteristics of a thermistor.

______

______

______

______

3. Compare the data of Tables 6-13A and B by noting whether each of the following measurements increased, decreased, or remained the same with an increase in temperature:
   (a) Voltage across R_1 (V_{R_1}).

   ______

   (b) Circuit current (I_T).

   ______

   (c) Voltage across thermistor (V_T).

   ______

   (d) Resistance of thermistor.

   ______

   (e) Circuit resistance.

   ______

Table 6-13A. Thermistor Characteristics (Room Temperature)

Source Voltage (V_S)	I_T	Voltage Across Thermistor (V_T)	Calculated Resistance of Thermistor (V_T/I_T)	V_{R_1}	Calculated Circuit Resistance (V_S/I_T)	Resistance Change of Thermistor
0.5						
1.0						
2.0						
3.0						
4.0						
5.0						
10						
15						
20						
25						
30						

(f) Resistance changes of thermistor.

____

4. In Step 2, why is it necessary to allow two minutes between source voltage changes?

____

____

5. In Step 6, what happened to the circuit current when the thermistor was heated?

____

____

____

____

Table 6-13B. Thermistor Characteristics (Heated)

V_S	I_T	V_T	Calculated Resistance of Thermistor	V_{R_1}	Calculated Circuit Resistance	Resistance Change of Thermistor
0.5						
1.0						
2.0						
3.0						
4.0						
5.0						
10						
15						
20						
25						
30						

Transducers—Thermistor Applications

Objective

In this laboratory activity, you will use a thermistor as a transducer to control a relay. When the thermistor is heated, it will cause the relay to change states. This application illustrates the temperature-sensitive characteristic of a thermistor.

Equipment

Thermistors—JA41J1, GE-X15.
Dc power source.
Transistors—2N3053 (3).
Relay—400 Ω coil.
Potentiometers—2500 Ω, 50,000 Ω, 100,000 Ω.
Resistors—680 Ω, 760 Ω.
Multifunction meter.
Cup of cold water.

Procedure

1. Construct the circuit shown in Fig. 6-14A.

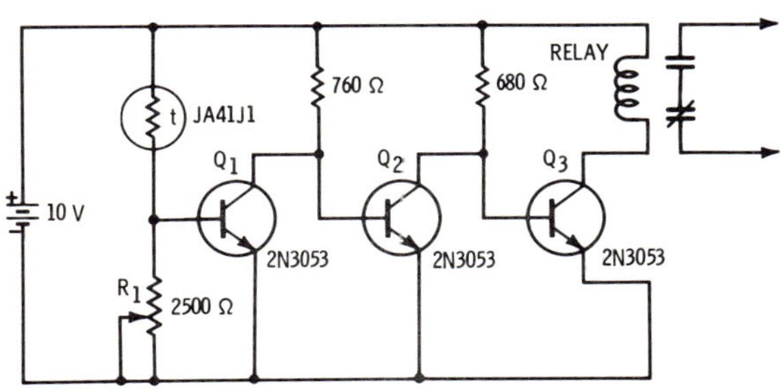

Fig. 6-14A. Thermistor transducer relay control circuit.

2. Adjust R_1 along its range, causing the relay to activate and drop out. This indicates that the circuit is operating properly.
3. Record the base current of Q_1 that is required to cause the relay to activate, as well as to drop out.

   Activate current = ______ milliamperes.

   Drop-out current = ______ milliamperes.
4. Adjust R_1 until the relay is almost ready to activate.
5. Grasp the thermistor with your fingers. Human body temperature is about 98° F. This small variation from room temperature should cause the relay to activate. If the relay does not change states, readjust R_1 until the relay operates properly.
6. Replace the thermistor and R_1 with a GE-X15 thermistor and a 100,000-ohm potentiometer connected as shown in Fig. 6-14B.

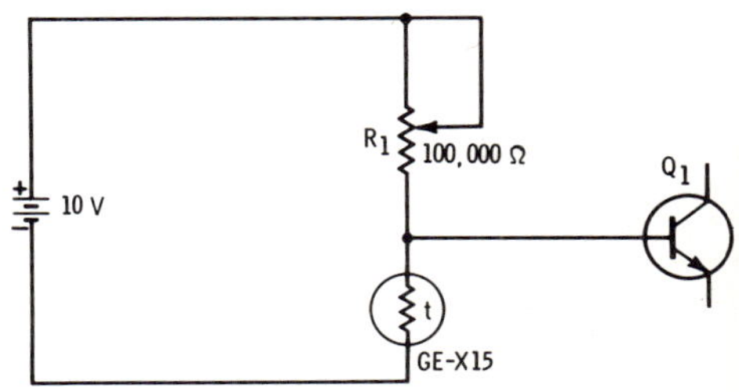

Fig. 6-14B. Alternate input for the circuit of Fig. 6-14A.

7. Adjust R_1 until the relay is activated but ready to drop out.
8. Immerse the GE-X15 probe in a cup of cold water. This should cause the relay to drop out. If not, readjust R_1, until the relay operates properly.
9. Remove the GE-X15 from the cup of water. This should allow the relay to activate. Adjusted properly, this circuit could be used as a liquid-level detector.

Analysis

1. What is the purpose of Q_1, Q_2, and Q_3 in the circuit of Fig. 6-14A?

2. In Step 3, why are the activate and drop-out currents not equal?

3. Why does the circuit of Step 6 cause the relay to respond differently than the circuit of Step 1?

____

____

____

4. What circuit modifications would have to be made to cause the circuit to detect hot water?

____

____

____

5. What are some industrial applications of thermistors?

____

____

____

____

____

____

____

____

6-15

Temperature Measurement

Objective

An important type of measurement used frequently to monitor industrial system operation is that of temperature. In this laboratory activity, you will construct a simple temperature measuring circuit using a GE-X15 thermistor bridge circuit.

Equipment

Dc power source.
Resistors—470 Ω, 1000 Ω (2).
Potentiometers—50 Ω, 1000 Ω, 5000 Ω.
Decade resistance box or appropriate resistor combinations.
Meter movement—0–100 μA, 500 Ω.
Thermistor—GE-X15.
Container of crushed ice.
Container of hot water.
Mercury thermometer.

Procedure

1. Construct the temperature measuring circuit as illustrated in Fig. 6-15A. The values of these components are critical to ensure accurate readings. Be sure that all selected values are as close as possible to those shown on the schematic.
2. The total resistance of the null branch of this bridge circuit must equal 1500 ohms. You should use a 0–100-μA, 500 Ω meter movement.
3. Place the thermistor in a cup of crushed ice with a small amount of water.
4. Adjust R_4 until the bridge is balanced at zero current. This represents a temperature of 0° Celsius.
5. Replace the thermistor with the calibration resistor supplied with the GE-X15. Adjust R_7 until the null meter indicates 50 microamperes. This represents a temperature of 50° C.

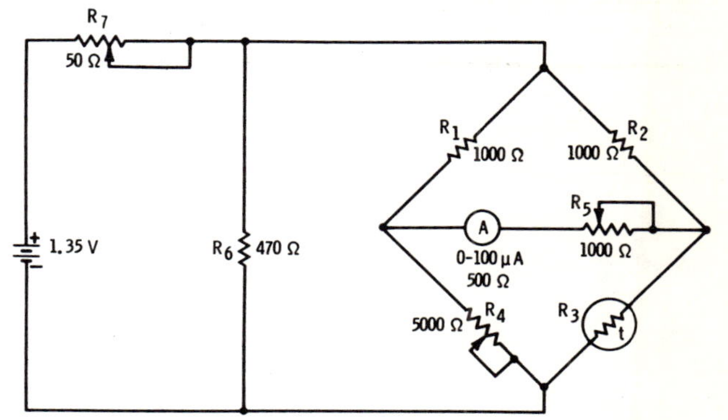

Fig. 6-15A. Temperature measuring circuit.

6. Return the thermistor to the circuit and return the calibrating resistor to its package.
7. Ideally, each microampere of current should now correspond to 1° C.
8. To determine the accuracy of your temperature measurement circuit, measure with a mercury thermometer the temperature of cold water, warm water, and the human body. Place the thermistor in the same environment and compare the readings. Complete Table 6-15A for each of these measurements.
9. Convert the Celsius readings of the null meter to Fahrenheit, Kelvin, and Rankine values.

Analysis

1. Analyze the operation of the thermistor thermometer circuit of this activity.

____

____

Table 6-15A. Temperature Measurements

Environment	Mercury Thermometer Reading (°F)	Null Meter (μA) (°C)	Experimental Thermometer Reading			
			°F	°C	K	°R
Hot water						
Cold water						
Human body						

2. How accurate were the readings obtained in Table 6-15A for the experimental circuit?

3. What are some industrial applications of this type of temperature measurement?

Force Through Liquids

Objective

One of the most significant functions of a mechanical power system is the transmission of force through a liquid. According to Pascal's law, pressure applied to a confined fluid is transmitted equally through the fluid in all directions, and is applied at right angles to the containing surfaces. Fluid does not have a specific shape, which lets it conform to the internal features of the container. Fluid also tends to resist compression when force is applied. These ideas are extremely important characteristics of an operating fluid system.

In this activity, two hydraulic cylinders are interconnected to form a simple closed-loop fluid system. Through this system you will be able to observe how force is transmitted through a fluid. You will also have an opportunity to test the compressibility of a hydraulic fluid.

Equipment

Double-acting cylinders—1⅛-inch bore (2).
Assortment of quick-disconnect transparent tubing pieces.

Procedure

1. Attach a length of flexible tubing to the port opposite the piston end of one of the 1⅛-inch bore doubling-acting hydraulic cylinders. Position the unit horizontally, with the ram or piston retracted into the cylinder.
2. Place the unattached end of the flexible tubing into a cup of hydraulic fluid. Manually force the piston in and out of the cylinder several times to charge it with oil and clear it of air.
3. Pull the piston all of the way out, thus fully charging it with oil. Set it aside momentarily.
4. Attach a length of flexible tubing to the port nearest the piston end of the second 1⅛-inch bore, double-acting hydraulic cylinder.
5. Using the same procedure outlined in Steps 2 and 3, clear the cylinder of air and charge it with oil. The piston should ultimately be fully retracted when it is charged.
6. Connect the unattached ends of the flexible tubing to the other cylinder as indicated in Fig. 7-1A.

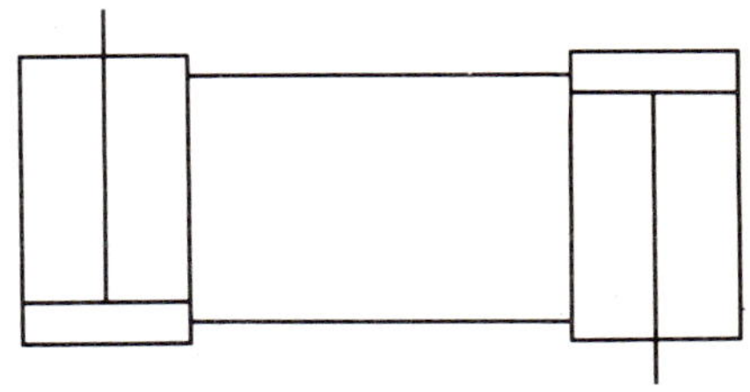

Fig. 7-1A. Hydraulic power test circuit.

7. Have a lab partner hold cylinder No. 1 while you hold cylinder No. 2. Pushing the ram of cylinder No. 2 down will transfer the force of cylinder No. 2 through the fluid to cylinder No. 1.
8. If transparent flexible tubing is used you should be able to see the direction of fluid flow with small trapped air bubbles.
9. Alter the position of the two cylinders and repeat the procedure. What does this indicate about a hydraulic system?

____

____

____

____

10. Adjust the position of the two pistons so that they are near the center of the cylinders. Apply force to both pistons. Ideally there should be no compression of the fluid. Any trapped air in the system will permit some degree of compressibility. How does your system respond?

____

____

____

____

11. Momentarily remove the flexible tubing from cylinder No. 1. Force the cylinder down approximately ¼ inch, then pull it back to its original position. Return the flexible tubing to the cylinder.

12. Adjust the position of the two cylinders near the center and repeat Step 10. How does the system respond?

13. Disconnect the flexible tubing from each cylinder and clear the cylinders of fluid. Wipe excess fluid off all components and return them to the storage area.

Analysis

1. What is meant by the term *compressibility?*

2. Explain how force is transferred through a fluid.

3. Describe the mechanical action of a double-acting cylinder.

A Static Hydraulic System

Objective

A hydraulic jack is undoubtedly one of the most common static fluid systems in operation today. With this system, a pump cylinder forces fluid into a ram cylinder through check valves. As a general rule, systems of this type are usually self-contained in a single housing.

In this activity, a simple hydraulic jack system will be constructed with discrete components. Through this activity you will have an opportunity to observe fluid flow through check valves that will permit an alternate direction of flow.

A hydraulic jack is often described as a static hydraulic system. This refers to the fact that the system develops its own operating pressure when energized by an outside mechanical force. A distinct mechanical advantage is developed by the two cylinders because of the differences in their size.

Equipment

Single-acting cylinder—1⅛-inch bore.
Single-acting cylinder—½-inch bore.
Check valves (2).
Flow control valve (needle type).
"T" connectors (2).
Pressure gauge—160 lb/in².
Assortment of quick-disconnect transparent tubing pieces.

Procedure

1. Assemble the hydraulic circuit of Fig. 7-2A. The 1⅛-inch bore cylinder will serve as the ram of the hydraulic jack, and the ½-inch bore cylinder will serve as the pump of the system. Mount both cylinders in an upright position.
2. Place the rams of both cylinders in the retracted position before attaching the flexible tubing to the cylinders, and close the flow control valve.
3. Pull the ram of the pump cylinder out of the cylinder, then force it down. One downward stroke of the pump cylinder causes the ram cylinder to move

_______ centimeters.

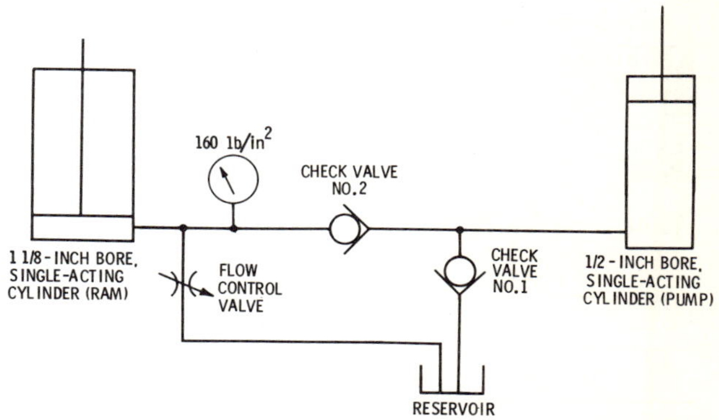

Fig. 7-2A. Hydraulic jack circuit.

4. How many strokes of the pump are needed to extend the ram?

   _______ strokes.
5. What pressure is indicated on the gauge?

   Pressure = _______ lb/in².
6. Open the flow control valve to drain the ram cylinder. Then close the valve to operate the system again.
7. Attach a weight of approximately ten pounds to the ram cylinder and repeat the operation of the jack.
8. What noticeable changes occur when the system is forced to do work?

   ____

   ____

   ____

   ____
9. Disconnect the system, drain the fluid from the cylinders and the tubing. Wipe any excess fluid from the components and return them to the storage cabinet.

Analysis

1. How much work was done by the ram cylinder in Step 7?

   ____

   ____

2. Describe the fluid circulation action that takes place during the operation of this system.

____

____

____

3. Does the pressure gauge change during the pumping action?

____

4. Why does this system have a mechanical advantage?

____

____

____

____

____

____

7-3

A Hydraulic Ram System

Objective

Hydraulic presses or rams play a very important role in industry today. Mechanical systems of this type are actuated by an electric-motor-driven pump unit that serves as the source of hydraulic energy. The transmission path of the system ultimately distributes energy to the load device, which does some form of mechanical work. Flow and directional control are achieved by special valves in the transmission line. An indicator is used to display system pressure. The hydraulic fluid is sent back to the pump reservoir through a return line. Systems of this type are classified as active units when compared with the static systems of activities 7-1 and 7-2.

In this activity, you will have an opportunity to identify some basic hydraulic components and their associated diagram symbols. You will then assemble these components into a functioning hydraulic ram system. Through this experience you will have an opportunity to observe the fluid flow through an active system change its direction by valve action. The basic operating principle of this simple hydraulic system has widespread utilization in industrial machinery today.

Equipment

Hydraulic power unit.
Flow control valve.
Two-position, four-way, lever-action control valve.
Doubling-acting cylinder—1⅛-inch bore.
Pressure gauge—160 lb/in².
"T" connector.
Assortment of quick-disconnect transparent tubing pieces.

Procedure

1. Select the necessary components to construct the hydraulic ram circuit of Fig. 7-3A.
2. Start the electric pump motor of the hydraulic power source. Slowly open the shutoff control valve of the circuit while observing the pressure gauge. Depending on the initial position of the four-way

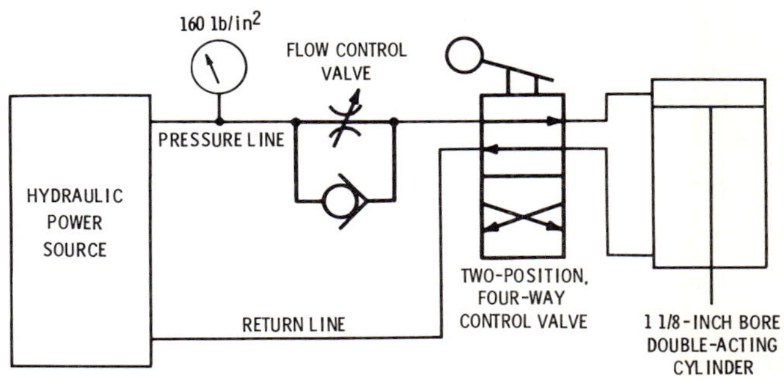

Fig. 7-3A. Hydraulic ram circuit.

control valve, the ram of the cylinder may extend itself slowly. Record the system pressure.

Pressure = ________ lb/in².

3. With the four-way control valve positioned as indicated in the circuit diagram, move the handle to the other position. What influence does this control action have on the ram of the double-acting cylinder?

____

____

____

4. Move the handle of the four-way control valve back to its original position to see the control action on the cylinder. Describe the action of the cylinder's ram.

____

____

____

5. As a general rule, the direction of fluid flow can be readily traced by the movement of tiny air bubbles in the transparent tubing. Switch the control valve to cause the ram to be extended and observe the direction of fluid flow. If the bubbles are not present, they can be introduced by closing the flow control valve and momentarily disconnecting the return line from the four-way control valve. Depress the check valve in the line and let a small amount of hydraulic fluid drain from the line. Then return the

line to the control valve and turn on the flow control valve.

6. Indicate the direction of fluid flow on the circuit diagram with solid-line arrows.
7. Switch the control valve to a position that will cause the ram of the cylinder to retract. Trace the fluid flow through the circuit. Make a sketch of the flow direction with broken-line arrows or with a different colored pen or pencil on the circuit diagram.
8. Turn off the electric pump motor of the hydraulic power source and disconnect the circuit. Wipe off any excess hydraulic fluid from the components and return them to their storage compartment.

Analysis

1. What are the primary system functions of the circuit constructed in this activity?

____

____

____

2. What specific component of the circuit is used to achieve each system function?

____

____

____

3. How could the ram of this circuit be used in an industrial application?

____

____

____

4. How does this system differ from the static systems of activities 7-1 and 7-2?

____

____

____

Hydraulic Power Source

Objective

The power source of an active hydraulic system is called a hydraulic pump. Primarily this component is responsible for producing a fluid flow that will ultimately develop pressure when it is forced to pass through system components. Fundamentally, a hydraulic pump must be in continuous operation whenever the system is operational. Through its action, hydraulic fluid is forced to move at a specific flow rate; which, in turn, causes a corresponding pressure to occur when resistance is encountered.

In this activity, you will investigate the basic components of a hydraulic power unit. This will include such things as the hydraulic pump, reservoir, filters, relief valve, pressure manifold, and return manifold. Pressure developed by this power source is variable, with a maximum rating of approximately 300 lb/in^2. For most experimental applications, the power source needs values of 100 lb/in^2 or less. Typical industrial hydraulic power sources may, however, call for values in the 1000- to 3000-lb/in^2 range.

Equipment

Hydraulic power source.
Pressure gauge—160 lb/in^2.
Assortment of quick disconnect transparent tubing pieces.
Measuring container—½ gal.
Open-ended disconnect tubing piece.

Procedure

Section A. Power Source Analysis

1. A basic hydraulic power source is needed to perform this activity. As a general rule, most of the components of this unit are permanently mounted and ready for operation.
2. Locate the electric motor of the unit and plug it in to the appropriate electrical outlet. Trace out the electrical circuit and locate the motor/pump on/off switch. In many units, the pump and motor may be mounted in a single housing.
3. Locate the system reservoir. Trace the fluid flow path from the reservoir to the inlet side of the hydraulic pump. Does your unit employ a fluid filter in this line?

   ____

4. Trace the fluid path from the outlet side of the pump to the pressure manifold. Are there any filters or controls in this part of the system?

   ____

5. Where is the pressure relief valve of your system located?

   ____

6. Turn on the electric motor/pump drive. Observe the pressure gauge and record the system pressure. If at all possible, see at what pressure the system returns fluid to the reservoir. This is determined by the pressure relief valve setting.
7. Connect a length of tubing to one of the pressure manifold valves. Run the other end of the tubing to an open tray or to the reservoir return line. Open the manifold valve momentarily until all of the air bubbles pass from the line. Does the pressure gauge change during this operation?

   ____

8. Turn off the electrical power to the pump drive motor. In the following space, make a sketch of your hydraulic power source pointing out specific component locations:

9. In the following space, draw a schematic diagram of the hydraulic power source using appropriate fluid power symbols:

Section B. Hydraulic Pump Delivery Rate

1. Attach a 160-lb/in² gauge to one of the pressure manifolds of the hydraulic power source. Connect a length of plastic tubing to the output side of a second manifold valve. Then connect the other end of the tubing to one of the return manifold connections.
2. Ready the hydraulic power source for operation by applying electricity and turning on the pump motor. Open the manifold valve attached to the pressure gauge. Record the system pressure. This valve also indicates the pressure setting or value of the pressure relief valve of the power source. If more than 100 lb/in² is indicated, ask the instructor to show you how to adjust the pressure limit.
3. Gradually open the manifold valve with the tubing attached. How does the system pressure gauge respond as this valve is opened?

___

4. Close the manifold valve and remove the plastic tubing with quick disconnects on each end. Replace it with a piece of tubing with a disconnect on only one end. The open end of the tubing will be used to fill a half-gallon measuring container to determine the pump delivery rate.
5. Open the manifold valve enough to maintain 100 lb/in² with the full output flowing through the tube. Fill the container and record the time it takes to pump ½ gallon. Run two trial tests at this pressure and average your results. Record your data in the following chart:

Pressure (lb/in²)	Half-Gallon Filling Time (seconds)	Pump Delivery Rate (gal/min)	Calculated Horsepower*
100			
80			

* Horsepower $= \dfrac{\text{gal/min} \times \text{lb/in}^2}{1714}$

6. Repeat the procedure outlined in Step 5 for a pressure setting of 80 lb/in².
7. Calculate the horsepower delivered by the pump for each pressure setting, and record the data in the chart.
8. Turn off the power unit and disconnect all circuit components. Wipe them free of excess fluid and return them to the storage cabinet.

Analysis

1. Explain the basic operation of the hydraulic power unit investigated in this activity.

___

___

___

2. Explain how the pressure of the system is altered.

___

___

___

3. What is the function of the pressure relief valve?

___

___

___

4. What type of displacement does the pump of the hydraulic power unit supply?

___

7-5

Hydraulic Gear Pumps

Objective

The pump of a hydraulic system is designed to provide an appropriate fluid flow that will ultimately develop system pressure. In an operating system, the pump accepts power from a drive motor and converts it into fluid flow.

Fundamentally, fluid flows into the inlet port of the pump and is expelled from the outlet port. After passing through the pump, fluid flow occurs at a faster rate. Ultimately, this increase in fluid flow causes an increase in system pressure when the flow encounters component resistance. Most hydraulic pumps displace a certain amount of fluid with each revolution of the impeller blade.

The gear pump used in this activity is a very popular method of changing rotary motion into fluid energy. Essentially this device contains two gears enclosed in a precision machined housing. Rotary motion from the power source is applied to the drive gear. Rotation of this gear causes the second or driven gear to turn, with their teeth meshing in the middle. The unmeshed gear teeth carry fluid from the inlet port to the center where the teeth mesh together and force the fluid to pass from the outlet port. Pumps of this type have an output capacity of from 0.5 to 1050 gallons per minute.

In this activity, you will construct a simple gear pump circuit and rotate the shaft of the pump manually to observe the direction of fluid flow. This will provide you with an opportunity to see how pump rotation and shaft speed influence the output of a hydraulic pump.

Equipment

Bidirectional, fixed-displacement, gear motor/pump.
Flow control valve.
Double-acting cylinder—1⅛-inch bore.
Assortment of transparent flexible tubing with disconnects.

Procedure

1. Connect the simple hydraulic pump circuit of Fig. 7-5A. The gear motor in this case will be used as a pump.
2. Rotate the shaft of the pump in a clockwise direction with one hand while holding it with the other hand. The tubing pieces should contain enough fluid to make the system operational.
3. Continue to rotate the shaft until the ram of the double-acting cylinder begins to move.

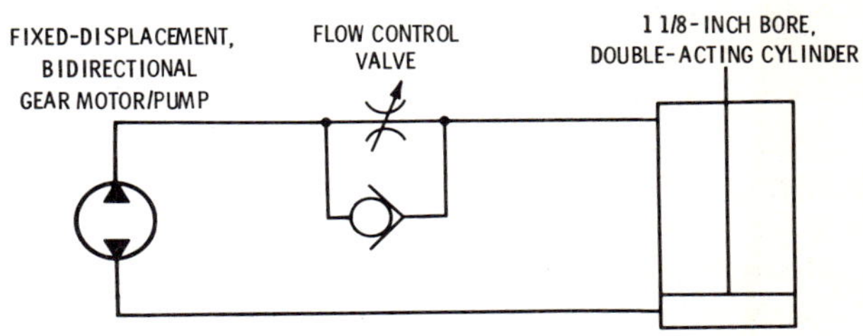

Fig. 7-5A. Hydraulic pump circuit.

4. Air trapped in the tubing pieces can be used to trace direction of fluid flow. Clockwise pump rotation causes the hydraulic fluid to flow in a ______ direction.
5. Rotate the shaft until the cylinder ram stops moving.
6. Rotate the shaft in a counterclockwise direction. Counterclockwise pump rotation causes hydraulic fluid to flow in a ______ direction.
7. Disconnect the circuit components and wipe off all excess fluid with a cloth. Return all components to the storage cabinet.

Analysis

1. How does pump speed of rotation influence the flow rate?

______

2. How does a gear pump produce fluid flow?

______

______

______

3. Explain the basic principle of fluid flow as achieved by a gear pump.

____

____

____

4. What would be a problem with a circuit of this type if it were connected to a drive motor?

____

____

____

7-6

Pressure Control Valves

Objective

When the pump of a hydraulic system is in operation, it supplies fluid to the system with each revolution of the drive motor. With the pump of a hydraulic system operating continuously, there needs to be an alternate flow path for fluid that is not being used by the system. Pressure relief valves placed in the power source are designed to provide an alternate flow path when the need arises.

In a strict sense, a pressure relief valve is a type of check valve that permits flow only when a certain pressure is reached. At reduced pressure, the valve checks the return flow and permits the fluid to pass into the system. At higher pressure levels it passes fluid back into the reservoir. In this application the relief valve is a safety device.

In this activity, you will have an opportunity to investigate the action of both a pressure relief valve and check valves more carefully. You will also have an opportunity to test the "cracking" pressure of the pressure relief valve. This refers to the release pressure that causes flow into the reservoir.

Equipment

Hydraulic power source.
Pressure relief valve.
Flow meter.
Flow control valve.
Assortment of quick-disconnect hoses.
Pressure gauge—160 lb/in².

Procedure

1. Select the components needed to assemble the circuit of Fig. 7-6A. Notice the two paths for the fluid flow in this circuit. This represents a typical power source system divider.
2. Adjust the pressure relief valve in the counterclockwise direction until it stops, then back off one turn of the valve control. Close the flow control valve completely.
3. Start the power unit and adjust the pressure relief valve to read 100 lb/in² of system pressure.

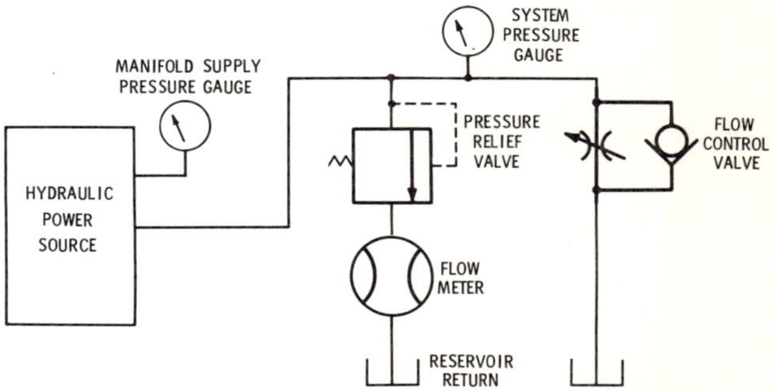

Fig. 7-6A. Hydraulic power source system divider circuit.

4. Open the flow control valve completely while observing system pressure. How does the pressure respond? Pressure with the flow control valve open is ________ lb/in².
5. Close the flow control valve again while observing the system pressure gauge. What does the flow meter do when the flow control valve is turned off?

____

6. Adjust the flow control valve to produce 100 lb/in² of system pressure. Then note the exact reading where flow first occurs through the flow meter. This is called the *cracking* pressure of the pressure relief valve.
7. Adjust the pressure relief valve one more turn as instructed in Step 2. Then adjust the flow control valve to 150 lb/in² of system pressure and determine the new cracking pressure of the pressure relief valve.
8. Turn off the power source and disconnect the circuit. Wipe all excess oil from the components and return them to the storage cabinet.

Analysis

1. Functionally, what does the pressure relief valve do in a system of this type?

____

____

____

2. Define cracking pressure.

3. What is the function of a flow control valve in a hydraulic circuit?

4. What accounts for a difference in manifold supply pressure and system pressure?

Hydraulic Directional Control

Objective

Directional control of a hydraulic system is primarily designed to alter the fluid flow in such a way that it starts, stops, or reverses the flow without altering system pressure appreciably. Valves that achieve this type of control are described as two-, three-, and four-way valves with different setting positions. Controls of this type may be actuated manually by pressure, with electricity, or by mechanical energy.

In this activity, you will construct a simple hydraulic motor control circuit and test the direction of flow through the circuit with a flow meter. As a general rule, this type of flow indicator will only respond to flow in a single direction. You will also be able to test the direction of flow with respect to different circuit positions in order to see how this device alters the flow direction. Through this experience you will become more familiar with the basic components of a hydraulic system and be able to trace out the flow path from manifold source to reservoir.

Equipment

Hydraulic power source.
Pressure relief valve.
Flow meter.
Three-position, four-way control valve.
Hydraulic gear motor.
Assortment of quick-disconnect hoses.

Procedure

1. Select the components needed to construct the hydraulic circuit of Fig. 7-7A. Note the location of the four test points. In this activity, you will be asked to insert the flow meter into each test point and observe the direction of flow. The flow meter will only respond when the flow is from the inlet port to the outlet port. Flow in the reverse direction will not produce an indication on the meter.
2. Close the pressure relief valve and back it off one complete turn. Then turn on the hydraulic power source and adjust the pressure relief valve for a manifold pressure of 100 lb/in^2 as indicated on the manifold supply gauge.

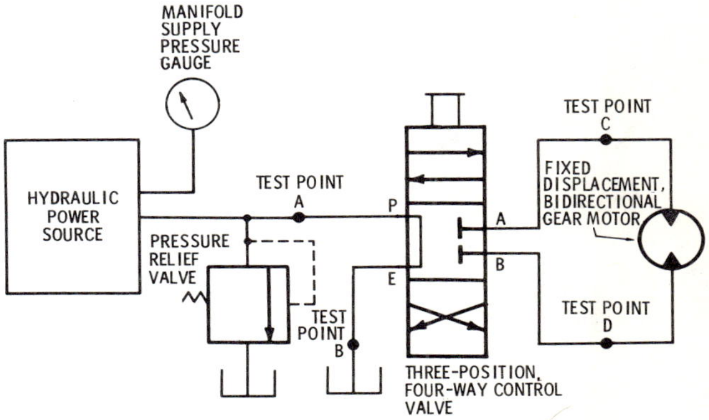

Fig. 7-7A. Hydraulic motor directional control circuit.

3. Momentarily turn off the power source and insert the flow meter into test point A. Then turn the power source on again.
4. Pull the position-shaft rod of the four-way valve to the fully extended position. Record the response of the flow meter and the motor.
5. Shift the control rod of the valve to the center position and record the flow meter and motor response.
6. Push the control rod into its most retracted position and note the flow meter and motor response.
7. Move the flow meter to test points B, C, and D, and repeat Steps 4, 5, and 6. Record the direction of flow and motor response for each situation.

8. In the following space, draw three separate schematic diagrams showing with arrows the direction of fluid flow for each of the three positions of the four-way valve.

9. Turn off the hydraulic power source and disconnect the circuit. Wipe all components free of excess fluid and return them to the storage cabinet.

Analysis

1. What do the letter designations P and E on the four-way valve denote?

____

____

2. What are the three motor operating conditions of the four-way valve?

____

____

____

3. Would the four-way valve be classified as a full-control or partial-control device?

____

7-8

Hydraulic Linear Actuators

Objective

The fundamental purpose of a hydraulic power system is to produce some form of mechanical work. The load part of the system is specifically responsible for achieving this function. The term *load*, however, is a rather broad term used to describe all components that consume power and do work. The term *actuator* is, therefore, used more commonly in hydraulic systems to describe specific components that achieve the work function.

Linear actuators of industrial hydraulic systems are primarily designed to lift, hold, compress, or clamp objects during different manufacturing processes. Hydraulic cylinders are the most significant linear actuator in operation today. These devices come in a variety of shapes and sizes that range from a fraction of an inch to several feet in diameter and length. The amount of force a cylinder can produce is determined by the surface area of the piston inside it and the pressure applied to it. Therefore, a cylinder with a larger surface area can produce more force than a smaller cylinder with the same pressure applied. Both single-acting and double-acting cylinders are included in the linear actuator classification.

In this activity, you will have an opportunity to construct a simple series and parallel linear actuator circuit and test the pressure and circuit operation under load and no-load conditions. Through this experience you will become more familiar with different system circuit configurations and observe some of the basic principles of series and parallel connected systems. Each type of circuit has a unique set of characteristics with respect to system pressure.

Equipment

Hydraulic power source.
Pressure relief valve.
Three-position, four-way control valve.
Pressure gauge—160 lb/in^2.
Double-acting cylinder—1⅛-inch bore (2).
"T" hose connectors (3).
Assortment of quick disconnect hoses.

Procedure

Section A. Series Linear Actuators

1. Select the appropriate components needed to construct the hydraulic circuit of Fig. 7-8A. Either a cylinder loading device or an actual weighted load needs to be attached to the cylinder in some steps of this activity.

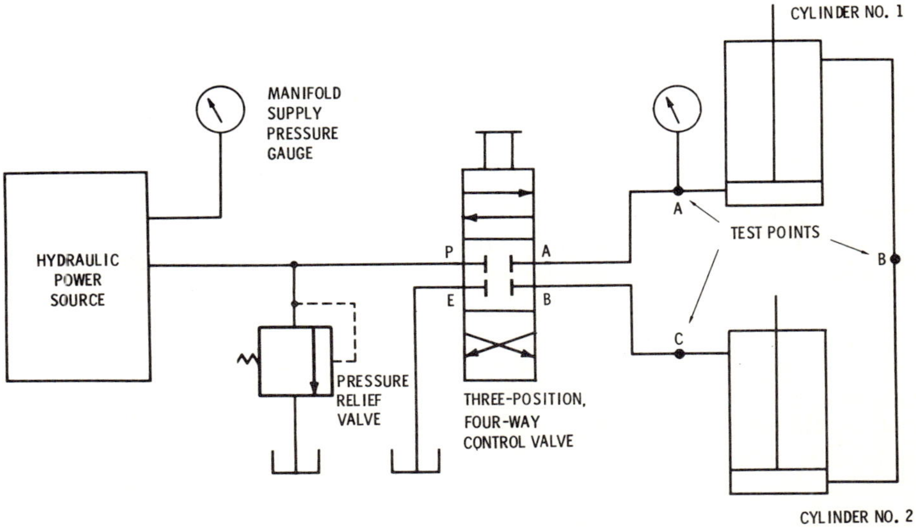

Fig. 7-8A. Hydraulic linear actuator circuit—actuators in series.

2. Turn on the power source and adjust the pressure relief valve to 120 lb/in^2.
3. Push the control lever of the four-way valve into the valve while observing the pressure at test point A. Make this reading while the cylinders are moving.

   Test point A pressure = ______ lb/in^2.
4. Move the four-way valve to the center or "off" position and move the pressure gauge to test point B. Repeat Step 3 and record the pressure.

   Test point B pressure = ______ lb/in^2.
5. Return the four-way valve to the center or "off" position and move the pressure gauge to test point C. Repeat Step 3 and record the pressure.

   Test point C pressure = ______ lb/in^2.
6. The pressure tests just taken represent the no-load test condition. If a cylinder loading device is available, connect it to cylinder No. 1 and adjust it to a heavy load. If a cylinder loading device is not available, attach a small platform plate to cylinder No. 1. Place 20 pounds of weight on the platform with the cylinder mounted in the vertical position.
7. Repeat Steps 3, 4, and 5 with the cylinder loaded. The pressure at test point C should be taken first since the gauge should be at this location from the previous test.
8. Make a chart showing the relationship of pressure between load and no-load conditions at each test point.
9. Increase the system pressure to 160 lb/in^2 and test the operation of the circuit in its loaded condition. How does this added pressure alter cylinder operation?

   ____
10. Reduce the system pressure to 80 lb/in^2 and test the operation of the circuit in its loaded condition. How does it respond to reduced pressure?

   ____
11. Turn the power source off and disconnect the circuit. Wipe all components free of fluid and return them to the storage cabinet.

Analysis

1. Why is this circuit described as a series linear actuator?

   ____

   ____

   ____
2. What influence does loading have on the pressure of a series linear actuator circuit?

   ____

   ____
3. What is a characteristic of the series circuit with respect to pressure throughout the system?

   ____

   ____
4. What general influence does system pressure have on the operational speed of a series linear actuator circuit?

   ____

   ____

Section B. Parallel Linear Actuators

1. Select the components needed to construct the hydraulic circuit of Fig. 7-8B.
2. Turn on the power source and adjust the pressure relief valve to 120 lb/in^2.
3. Push the control lever of the four-way valve into the valve while observing the pressure at test point A.

   Test point A pressure = ______ lb/in^2.

   Now pull the control lever completely out while observing the pressure. Make a pressure reading while the cylinders are in motion.

   Pressure = ______ lb/in^2.

   Does the flow direction influence the pressure?

   ____
4. Place the four-way control valve in the center or "off" position. Place the pressure gauge at test points B and C and repeat the procedure outlined in Step 3.

   Test point B pressure = ______ lb/in^2.

   Test point C pressure = ______ lb/in^2.

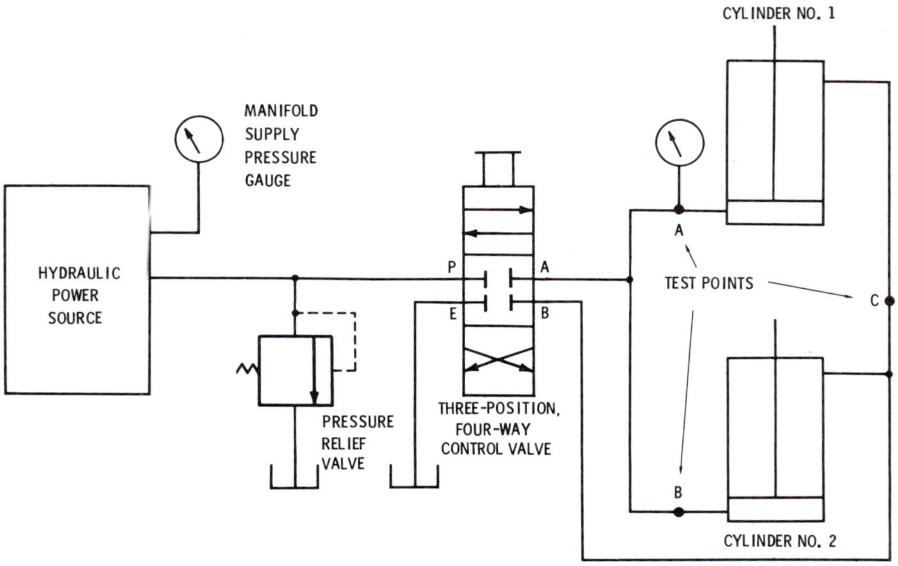

Fig. 7-8B. Hydraulic linear actuator circuit—actuators in parallel.

5. These tests represent the no-load condition of the linear actuator. If a cylinder loading device is available, connect it to cylinder No. 1 and adjust it to the heavy load condition. If this device is not available, attach a small platform plate to cylinder No. 1. Place 20 pounds of weight on the platform with the cylinder mounted in the vertical position.
6. Repeat Steps 3, 4, and 5 with the cylinder loaded. The pressure at point C should be taken first since the gauge should be at this location from the previous test.
8. Make a chart showing the relationship between load and no-load pressure conditions at each test point.

9. Increase the system pressure to 160 lb/in^2 and test the operation of the circuit in its loaded condition. How does this added pressure alter the operation of the cylinders?

____

10. Reduce the system pressure to 70 lb/in^2 and test the operation of the circuit in its loaded condition. How does it respond to reduced pressure?

____

11. Turn off the power source and disconnect the circuit. Wipe all components free of excess fluid and return them to the storage cabinet.

Analysis

1. Why is this circuit classified as a parallel linear actuator?

____

____

____

2. What influence does loading have on the system pressure of a parallel linear actuator system?

____

____

____

3. What is a characteristic of the parallel circuit with respect to pressure throughout the system?

____

____

____

4. What general influence does system pressure have on the operational speed of a linear actuator?

____

____

____

7-9

Hydraulic Rotary Actuators

Objective

Hydraulic rotary actuators are designed to produce an appropriate form of rotary motion in either direction. Fluid under pressure applied to the side of this device flows through the rotating member and causes it to move before it is expelled from the outlet side. Counterclockwise rotation is achieved by reversing the direction of fluid flow. These devices are used in industry today to open and close gates, for indexing operations, and in lifting operations. As a general rule, the speed of hydraulic rotary actuators is rather slow compared to electric actuators, but they develop a great deal of torque.

In this activity, a hydraulic gear motor/pump device will be used as a rotary actuator. The turning ability of this device is a measure of the work the device is designed to do. Torque is used to describe the developed amount of work. It is determined by the developed force multiplied by the radius of the rotating arm. You will also have an opportunity to observe the influence of pressure and loading upon the rotational speed of the actuator. Rotary actuators represent a very important hydraulic system load device.

Equipment

Hydraulic power source.
Pressure relief valve.
Three-position, four-way control valve.
Flow meter.
Pressure gauge—160 lb/in².
Bidirectional, fixed-displacement, gear motor/pump.
"T" connector.
Assortment of quick-disconnect hoses.
Tachometer.

Procedure

1. Select the appropriate components needed to construct the hydraulic circuit of Fig. 7-9A.
2. Turn on the power source and adjust the pressure relief valve for 120 lb/in².
3. Push the control lever of the four-way valve into the valve while observing the direction of motor rotation.

   Direction of rotation is ______________ (clockwise or counterclockwise).

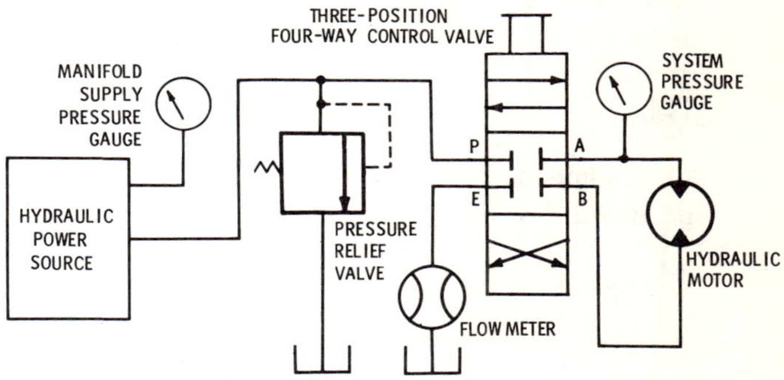

Fig. 7-9A. Hydraulic rotary actuator circuit.

4. Pull the valve control lever to its most extended position.

   Direction of rotation is ______________.
5. Make this test again and note the system pressure needed to produce rotation.
6. The flow meter indicates ________ gal/min.
7. If a tachometer is available, measure and record the rotational speed of the motor. You can also measure speed when it is slow enough by marking the shaft and counting the number of revolutions per minute.

   Rotational speed = ________ r/min.
9. This represents the unloaded speed and pressure of the motor. Increase the pressure to 160 lb/in². Record the influence that this has on speed and flow rate.

   Speed = ________ r/min.

   Flow rate = ________ gal/min.
10. Reduce the pressure to 80 lb/in² and record the influence that this has on speed and flow rate.

    Speed = ________ r/min.

    Flow rate = ________ gal/min.
11. Attach a rotary load of some type to the mounted hydraulic motor.

12. Repeat Steps 2 through 10 with a substantial load on the motor. Record your findings in the following chart:

Pressure (lb/in²)	Speed (r/min)	Flow Rate (gal/min)
80		
120		
160		

13. Turn off the power source and disconnect the circuit. Wipe all components free of excess hydraulic fluid and return them to the storage cabinet.

Analysis

1. What influence does system pressure have on motor speed and flow rate?

2. What influence does loading have on pressure and speed?

3. Explain how a hydraulic motor produces rotation.

4. What does the term *displacement* refer to in hydraulic motors?

7-10

Compressibility of Air

Objective

Undoubtedly the most significant difference in a hydraulic and a pneumatic power system is in the compressibility of air with respect to fluid. Pneumatic systems, for example, are designed to compress air into smaller volumes and increase its pressure. Compressed air can then be stored in a receiver tank and released when the demand for it arises. This of course removes the necessity of continuous pump operation that is characteristic of a hydraulic system.

The pressure associated with a pneumatic system can be expressed in a number of ways. Barometric, or absolute, pressure, for example, is 14.7 lb/in^2 at sea level. Gauge pressure is the difference in absolute pressure and an unknown system pressure. A gauge reading of 100 lb/in^2 represents an absolute pressure of 114.7 lb/in^2. Pressure below atmospheric pressure is commonly considered as a negative pressure or as a vacuum.

When a certain volume of air or gas is squeezed together, it causes a corresponding increase in pressure. Essentially, a decrease in container volume causes a corresponding increase in the density of the molecules. This, in turn, increases the rate at which the molecules strike the walls of the container, which causes a corresponding increase in pressure. Likewise, when the volume of a container is increased, it causes the gas molecules to spread apart. This, in turn, causes a corresponding reduction in total container pressure. This relationship is normally expressed by the ratio of: $P_1 \times V_1 = P_2 \times V_2$. P_1 and V_1 refer to the pressure and volume of the original container, while P_2 and V_2 refer to its reduced or expanded condition.

In this activity, you will have an opportunity to construct a simple pneumatic system and test the compressibility of air while observing the changes in pressure. The internal, or "blind-end," volume of a cylinder is based on the area of the piston times the length of the stroke. Area of the cylinder, which is measured in square inches or square centimeters, is determined by the formula: Area = $D^2/4$. The blind-end volume of a cylinder is, therefore, determined by the cylinder area multiplied by the stroke length, and is expressed in cubic inches or cubic centimeters.

Equipment

Air supply source—0–100 lb/in^2.
Check valve.
System pressure gauge—100 lb/in^2.
Double-acting pneumatic cylinder—1⅛-inch bore.
Assortment of quick-disconnect hoses.

Procedure

1. Construct the pneumatic circuit of Fig. 7-10A. Note that this circuit includes several items that are part of the primary air supply. These items, as a rule, are optional and may not be included in the supply

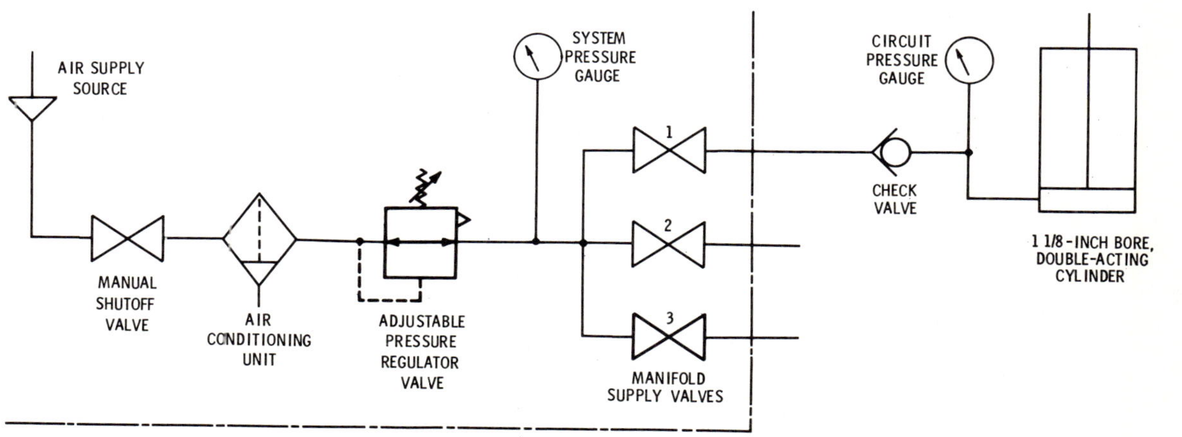

Fig. 7-10A. Pneumatic power test circuit.

feeding each work area. The part of the circuit that we are primarily concerned with begins at the manifold supply valve.

2. Close manifold supply valves 1, 2, and 3 and open the manual shutoff valve of the air source. Adjust the system pressure regulator to read 15 lb/in^2 on the system pressure gauge.
3. Measure the length of the cylinder shaft when it is fully retracted.

   Length = ______ inches.
4. Slowly open manifold supply valve 1 while observing the action of the gauge and the cylinder. If the circuit does not respond, turn off the manifold valve and check the flow direction of the check valve.
5. When the cylinder ram is fully extended, measure and record the circuit pressure.

   Pressure = ______ lb/in^2.

   Measure and record the extended length of the cylinder ram.

   Extended length = ______.

   Calculate the blind-end volume of the cylinder.

   Cylinder volume = ______.
6. Manually force the piston back into the cylinder one-half of its extended distance while observing the circuit pressure gauge.

   Pressure = ______ lb/in^2.

   Measure the extended rod length and calculate its blind-end volume.
7. Release the cylinder piston and let it extend again.
8. Adjust the supply regulator to 25 lb/in^2 and repeat Steps 3, 4, 5, and 6. How does this action differ from the preceding steps?

   ____

   ____

   ____
9. Close the manifold supply valve and the main air supply shutoff valve. Disconnect the circuit and return all components to the storage cabinet.

Analysis

1. Describe the relationship between blind-end cylinder volume and pressure.

   ____

   ____

   ____
2. What is the absolute pressure of Step 5?

   ____
3. As pressure increases how does its compressibility change?

   ____
4. Why does pneumatic pressure increase when the volume of a container is reduced?

   ____

   ____

   ____

7-11

Pneumatic Source Analysis

Objective

The air supply source of a pneumatic system is an extremely important part of the entire pneumatic system. As a general rule, it is energized by an electric motor that drives a compressor of some type. The compressor, as the name implies, squeezes certain volumes of air together and forces it into the receiving tank. The receiving tank then holds this compressed air until it is to be utilized by the system. Depending on the demand of the system for air, the electric drive motor of the compressor may only be called upon to operate periodically when the air supply diminishes. The basic principle of air compression and storage represents a unique characteristic of a functional pneumatic system.

In this activity, you will have an opportunity to study and observe the supply system that feeds air to your work station. This part of the system typically includes a motor-driven compressor, a receiver tank, air conditioning equipment, pressure control valves, regulators, a distribution manifold, and pressure limit switches that respond to tank pressure changes. The air pressure producing capacity of the compressor is primarily designed to meet the demands of the total system being supplied. Typical industrial pneumatic systems operate within a range of 80 to 200 lb/in^2 gauge pressure.

Equipment

Operating air supply source.

Procedure

1. With the assistance of the instructor locate the primary air source of the system supplying your work station.
2. Describe the basic type of compressor employed by the system.

____

____

____

____

3. What are the essential parts of this specific air source?

____

____

____

____

4. Make a symbol diagram of the complete air source.

5. What is the maximum receiver tank pressure developed by the system?

____

6. At what tank pressure does the compressor motor turn on to build up receiver tank pressure?

____

7. Describe the distribution path from the receiver tank to your work station.

____

____

____

8. Describe some periodic maintenance procedures that must be performed on the compressor and air source components of this part of the system.

____

____

____

____

Analysis

1. What are some of the primary differences in a hydraulic source and a pneumatic system source?

____

____

____

2. What are some of the basic compressor types in operation today?

____

____

3. How does compressing air increase its pressure?

____

____

____

7-12

Pneumatic Pressure and Flow

Objective

When air passes through a pneumatic system, it normally encounters an opposition that tends to reduce or lower the pressure between different locations. This lowering of pressure is primarily due to a form of friction that air encounters as it flows through the system. When air flow is reduced or stopped, it causes a change in friction. In a closed system without an air flow, the pressure is equalized throughout. When the flow of air increases, there is a corresponding increase in pressure drop. The length of the transmission flow path, the diameter of the tubing, number of and type of direction changes, and path restrictions all have some influence on the amount of pressure drop developed within a system.

In this activity, you will have an opportunity to construct a simple pneumatic system and alter the flow of air to see how it influences system pressure. You will be able to observe both an equalized system condition and the pressure drop of a heavy flow circuit. These conditions are very important considerations in the operation of a pneumatic system.

Equipment

Air supply source.
Flow control valve.
Pressure gauge—0–100 lb/in^2.
Manual shutoff valve.
Muffler.
"T" connector.
Assortment of quick-disconnect hoses.

Procedure

1. Construct the pneumatic circuit of Fig. 7-12A.
2. Close the manifold supply valves and turn on the manual shutoff valve. Adjust the pressure regulator valve to 15 lb/in^2 of system pressure. Close the circuit flow control valve and the circuit shutoff valve, then slowly open the manifold supply valve. The circuit pressure gauge should read zero at this time if everything is adjusted properly. If not, check the two circuit valves and make certain they are off.
3. Gradually open the flow control valve to produce a reading on the circuit pressure gauge. How does this reading compare with the system pressure?

____

4. Open the circuit manual shutoff valve. How do the two gauge pressures compare?

____

5. Turn off the circuit shutoff valve and add approximately six feet of hose between the muffler and the shutoff valve.
6. Turn on the manual shutoff valve again and observe

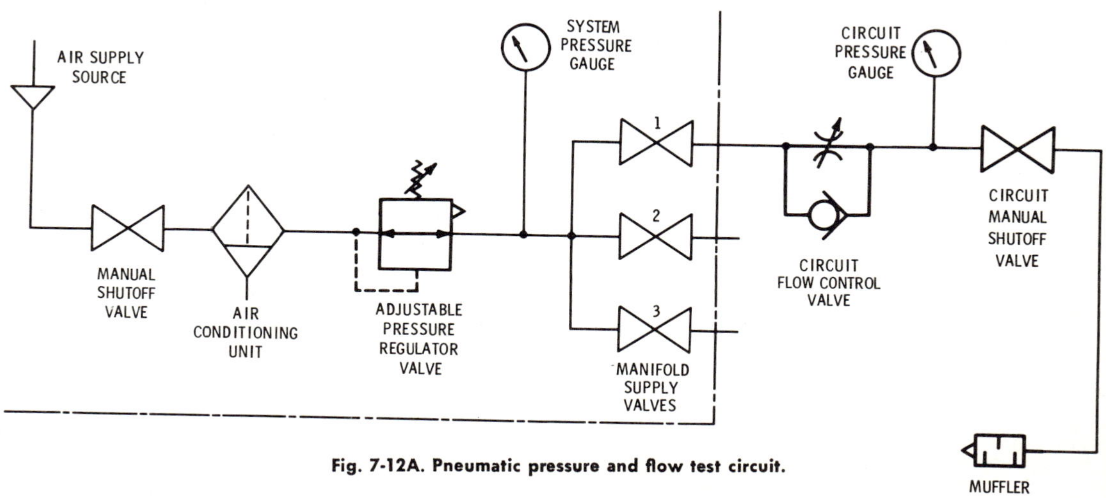

Fig. 7-12A. Pneumatic pressure and flow test circuit.

circuit pressure. What influence does the additional hose have on the pressure reading?

____

7. Turn off the circuit shutoff valve and remove the muffler with the six-foot length of hose remaining.
8. Turn on the shutoff valve and observe the pressure reading. How does this compare with the value of Step 6?

____

9. Turn off the manifold supply valve and the manual shutoff valve of the air supply. Disconnect the circuit and return all components to the storage cabinet.

Analysis

1. What causes the pressure of a pneumatic circuit to drop at different locations?

____

____

____

2. What is meant by the term *system resistance*?

____

____

____

____

____

____

____

3. What are some of the things that alter the resistance to the flow of a pneumatic system?

____

____

____

____

____

____

____

7-13

Pneumatic Linear Actuators

Objective

Pneumatic linear actuators are primarily used in industrial applications to lift, compress, hold, or position objects during different manufacturing processes. In order for this action to be achieved, air is fed into a chamber under pressure. A piston or ram in the chamber is then forced to move depending on the force developed by the applied pressure. Pneumatic cylinders range in size from a fraction of an inch to several inches in diameter. Both single-acting and double-acting cylinders are used to achieve linear mechanical action in pneumatic circuits today.

In this activity, you will have an opportunity to construct a simple linear actuator circuit to test the lifting capabilities of air. You will have an opportunity to observe this condition of operation with no load and under a loaded condition. Through this experience you will become more familiar with basic pneumatic circuits and their operation. You will also have an opportunity to employ a flow control valve, which serves as a throttle that alters cylinder operating speed. This circuit modification is very important in many industrial applications today.

Equipment

Air supply source.
Three-position, four-way control valve.
Single-acting cylinder—1⅛-inch bore.
Double-acting cylinder—1⅛-inch bore.
"T" connectors—(2).
Pressure gauges—0–100 lb/in² (2).
Flow control valve.
Assortment of quick-disconnect hoses.
Platform plate and 10-pound weight.

Procedure

Section A. Single-Acting Linear Actuators

1. Construct the linear actuator circuit of Fig. 7-13A. Note that the four-way valve has a plug attached to part of the valve that is not used.
2. Turn off the manifold supply valve. Energize the air supply source and adjust the pressure regulator valve for 15 lb/in².
3. Turn on the manifold supply valve and switch the four-way valve to all three positions. Describe the action of the cylinder.

____

____

____

4. Place a weight on the cylinder or load it by holding it down with your hands. Run through the switching sequence again with an increased load. How does this influence the action of the cylinder?

____

5. Increase the system pressure to 40 lb/in² and repeat Steps 3 and 4. How does increased pressure alter the operation of the cylinder?

____

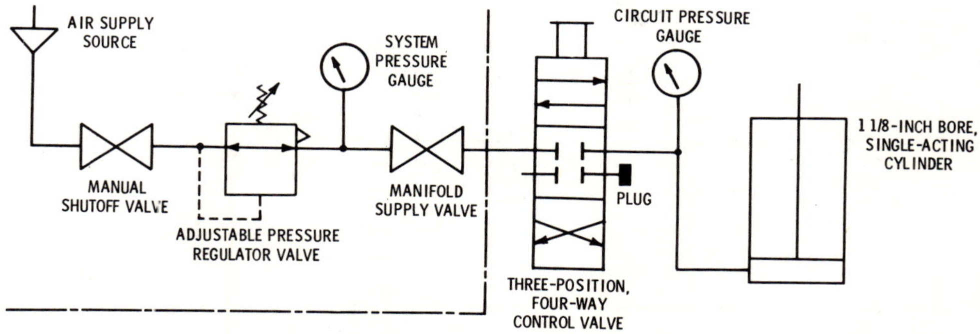

Fig. 7-13A. Single-acting pneumatic linear actuator circuit.

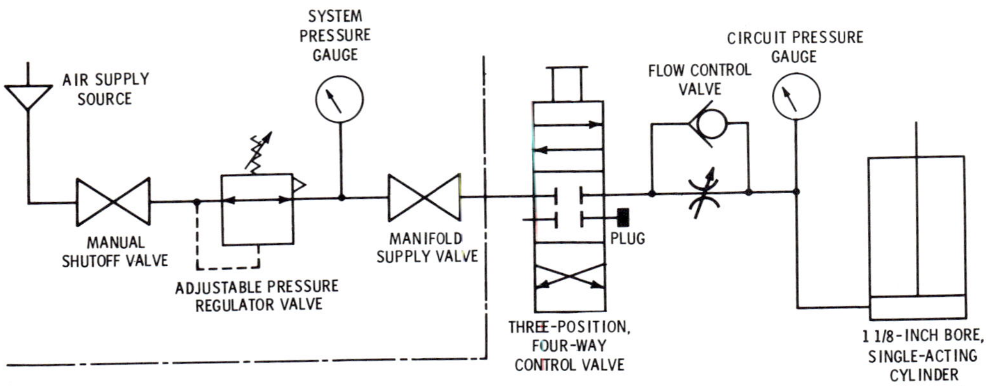

Fig. 7-13B. Flow control valve added to circuit of Fig. 7-13A.

6. Turn off the manifold supply valve and insert the flow control valve as indicated in Fig. 7-13B.
7. Turn on the manifold supply valve and switch the four-way valve to all three positions. In the retracting position, the flow control valve should be altered to regulate the retracting speed. This type of valve serves as a throttle to regulate cylinder retraction.
8. Turn off the manifold supply valve and reverse the flow direction of the flow control valve.
9. Turn on the manifold supply valve and switch the four-way valve through its three positions. How does this alteration change the operation of the circuit?

____

____

____

10. Turn off the manifold supply valve and the air supply shutoff valve. Disconnect the circuit and return all parts to the storage cabinet.

Analysis

1. What is an industrial application of a pneumatic circuit of the type investigated in this activity?

____

____

____

2. What is a unique difference in this circuit operation and that of a similar hydraulic circuit?

____

____

____

3. Explain the operation of the throttling valve technique of cylinder speed control.

____

____

____

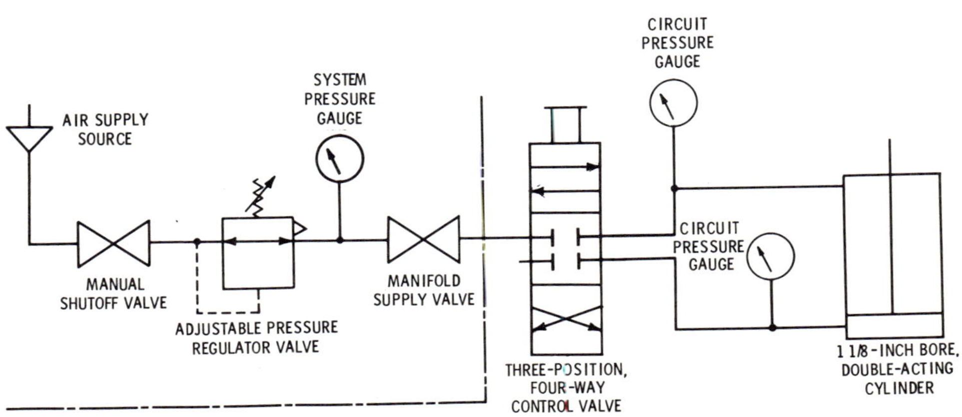

Fig. 7-13C. Double-acting pneumatic linear actuator circuit.

Section B. Double-Acting Linear Actuators

1. Construct the double-acting linear actuator circuit of Fig. 7-13C.
2. Turn off the manifold supply valve of the air supply and energize the source. Adjust the regulator pressure to 40 lb/in^2.
3. Turn on the manifold supply valve and switch the four-way valve to all three positions. Make a note showing the extended and retracted positions of the four-way valve. How do the pressure readings respond during operation of the cylinder?

____

____

____

4. Attach a platform to the top of the cylinder with the cylinder extending upward. Place a 10-pound weight on the platform and repeat Step 3. What influence does an increased load have on the operation of the circuit?

____

____

5. Turn off the manifold supply valve and place a flow control valve in the top cylinder line of the circuit. Connect as in Fig. 7-13B so that the free flow is from the four-way valve to the cylinder with the throttled flow from the cylinder to the four-way valve.
6. Turn on the manifold supply valve and switch the four-way valve to all three positions. In the retracting position, the flow control valve should be adjusted to produce a slow retracting action.
7. Turn off the manifold supply valve and switch the location of the flow control valve to the bottom line.
8. Turn on the manifold supply valve and switch the four-way valve through its three positions. How does the control valve alter the operation of the cylinder in this circuit location?

____

____

9. If two flow control valves are available, you can place one in each cylinder line and alter the operational speed of the valve in both the retracting and extending positions.
10. Turn off the manifold supply valve and the air supply shutoff valve. Disconnect the circuit and return all parts to the storage cabinet.

Analysis

1. Make a circuit diagram for each of the three operating positions of the four-way valve with arrows showing the path of the air flow.

2. Why is the doubling-acting linear actuator considered to have better control than a single-acting cylinder?

____

____

____

3. Why does the four-way valve not require modification in this circuit as it did in Fig. 7-13A?

____

____

____

Pneumatic Rotary Actuators

Objective

Pneumatic rotary actuators are designed to produce rotary motion from air pressure. Air under pressure is applied to the rotating member, causing it to move. The air is then expelled from the outlet side. As a general rule, reversing the direction of air flow causes the motor to rotate in the opposite direction. Pneumatic motors are lightweight, compact, and have a wide range of variable speed characteristics when compared with electric or hydraulic motors. These characteristics are particularly well suited for hand power tools, hoists, mixers, and in explosion-proof applications.

The horsepower and speed of a pneumatic motor can readily be changed by throttling the applied air flow. As the load applied to a pneumatic motor is increased, it normally causes a reduction in operational speed. This behavior is similar to that of an electric motor. With an increased load to an electric motor, however, the speed of rotation must quickly return to the normal running speed or it will be damaged by excessive heat. Excessive loading of a pneumatic motor will not adversely damage the motor because the air flow has a significant cooling effect. It can be run at alternate speeds without adverse damage to the motor for prolonged periods of time.

Pneumatic motors are energized by air, which means that they can be used in potentially explosive environments. This characteristic is extremely important in many industrial applications today.

The motor used in this activity produces up to 1/3 horsepower and operates from 500 to 10,000 r/min. It is primarily designed to respond to pressures below 100 lb/in². The amount of air flow needed to operate this motor is quite high, however. As a rule, the overall operating efficiency of a pneumatic motor is less than 20% for continuous operation. Air motors are, therefore, rarely used when operational efficiency is important. They are used, however, where the variable speed characteristic or explosion-proof advantage outweighs efficiency.

In this activity, you will construct a simple pneumatic motor circuit and test the loading effect on applied pressure and speed. You will then test the direction of rotation and see if loading and speed are altered. Through this activity, you will become more familiar with the operation of a pneumatic motor and see how rotary actuators can be harnessed to do work.

Equipment

Air supply source.
Flow control valve.
Three-position, four-way control valve.
Reversible pneumatic motor—Gast No. 1AM vane type.
Muffler.
Pressure gauges—0–100 lb/in² (2).

Procedure

Section A. Rotary Actuator Characteristics

1. Construct the rotary actuator circuit of Fig. 7-14A.
2. Turn off the manifold supply valve, then energize the air supply. Adjust the pressure regulator valve for 30 lb/in² of system pressure.
3. Close the flow control valve, then open the manifold supply valve. Slowly open the flow control

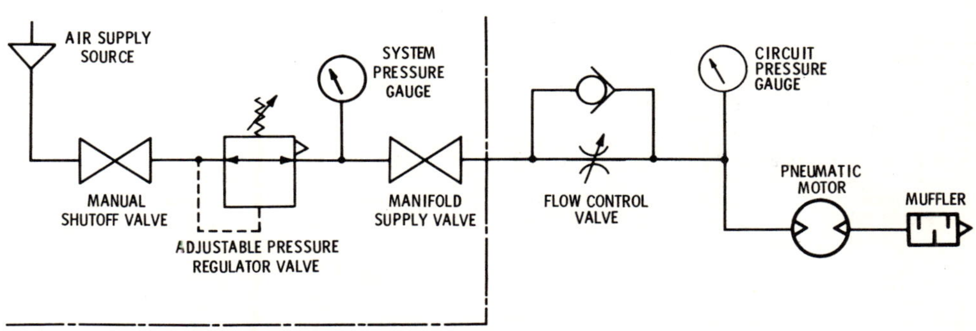

Fig. 7-14A. Pneumatic rotary actuator circuit.

valve while observing the circuit pressure gauge. At what pressure does rotation occur?

Rotation occurs at ______ lb/in².

4. Note the direction of motor rotation.

Direction of rotation is ____________ (clockwise or counterclockwise).

5. Gradually increase the flow while observing the circuit pressure. What is the maximum pressure observed?

Maximum pressure = ______ lb/in².
Note that when the pressure increases, the speed increases accordingly.

6. Connect a tachometer to the motor shaft and test the no-load speed.

Speed = ______ r/min.

7. Load the motor by carefully wedging a piece of wood between the rotating shaft and the base of the motor. How does loading influence speed and circuit pressure?

____

____

8. Momentarily turn off the manifold supply valve and wait for the motor to stop rotation.
9. Adjust the pressure regulator valve to 70 lb/in² of system pressure. Close the flow control valve.
10. Turn on the manifold supply valve and gradually turn on the flow control valve. The motor should be loaded with the wooden wedge again. Increase the flow until rotation occurs. What circuit pressure is needed to produce initial rotation of the motor under a loaded condition?

Pressure = ______ lb/in².
How does this compare with the unloaded condition of Step 3?

____

____

11. Turn off the manifold supply valve and turn off the air supply source. Disconnect the circuit and return all components to the storage cabinet.

Analysis

1. What influence does loading have on motor starting torque and the applied starting pressure?

____

____

____

2. How does loading alter the rotational speed of a pneumatic motor?

____

____

____

3. Of what significant value would a pneumatic motor be in industry?

____

____

____

Section B. Rotary Actuator Directional Control

1. Construct the rotary actuator circuit of Fig. 7-14B.
2. Turn off the manifold supply valve and energize the air supply source. Adjust the pressure regulator valve to 70 lb/in² of system pressure.
3. Turn off the flow control valve, then turn on the manifold supply valve. Switch the four-way directional control valve to each of the three positions. Then switch the four-way valve to one of its extreme positions.
4. Carefully adjust the flow control valve to produce a slow rotation of the motor. Make a note indicating the direction of rotation produced by this switch position.

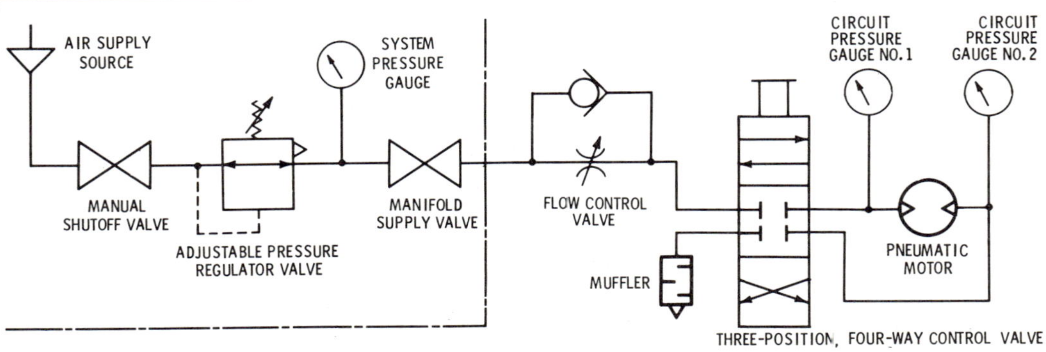

Fig. 7-14B. Pneumatic rotary actuator directional control circuit.

5. Record the pressure indicated by gauges 1 and 2.

   Gauge No. 1 = ______ lb/in².

   Gauge No. 2 = ______ lb/in².
6. Switch the four-way control valve to the center or "off" position. Wait momentarily until rotation stops. Then switch the four-way control valve to the other extreme position. Make a note indicating the direction of rotation produced by this switch position.
7. Record the pressure indicated by gauges 1 and 2.

   Gauge No. 1 = ______ lb/in².

   Gauge No. 2 = ______ lb/in².
8. Increase the speed of rotation by opening the flow control valve. With a tachometer, measure the rotational speed. (Note: Some tachometers will only indicate speed in one rotational direction. Test the tachometer before making this measurement.)

   Speed = ______ r/min.
9. Switch the four-way control valve to the "off" position until the motor stops. Then switch it to produce rotation in the opposite direction. Observe the note in Step 8.

   Speed = ______ r/min.
   Does the motor run equally well in either direction?

   ____
10. If time permits, repeat Steps 1–9 after wedging a piece of wood between the motor shaft and its base as a loading device.
11. Turn off the manifold supply valve and the supply shutoff valve. Disconnect the circuit and return all parts to the storage cabinet.

Analysis

1. How would the pressure gauges of this circuit respond if the muffler were removed?

   ____

   ____
2. What are some of the common types of pneumatic motors in operation today?

   ____

   ____

   ____
3. What kind of pneumatic motor is used in this activity?

   ____
4. How does a pneumatic motor compare with an electric motor?

   ____

   ____

   ____

7-15

Electromechanical Relays

Objective

The electromagnetic relay is undoubtedly the most widely used electromechanical device in industry today. Devices of this type are designed to actuate or control load current from an independent source by electromagnetic action. As a general rule, the contact points are electrically insulated from the relay coil and are related only to the electromagnetic field produced by the coil. The moving part of the relay is commonly called the *armature*. Energizing the relay coil will cause the armature to move physically due to its attraction to the electromagnetic field. This mechanical action then causes the contacts to be engaged or disengaged according to the design of the relay. A large number of general-purpose and specialized relays with numerous switching combinations are available today.

In this activity, you will have an opportunity to test the "pull-in" and "drop-out" current characteristics of a relay. The procedure outlined in this activity is commonly used to test the characteristics of an unknown relay or to determine if a particular relay is responding as it should. Through this activity, you will gain some insight into the basic operation of an electromagnetic relay.

Equipment

Relay—12-V dc, 125-Ω coil (Guardian Electric No. 1335-2C-12D or equivalent).
Dc power supply—0-20 V.
Ac power source—6.3 V.
Dc ammeter.
Multifunction meter.
Lamp—No. 47.
Toggle or push-button switches—spst (2).
Diode—1N4004.
Capacitor—4 μF, 25 V dcw.

Procedure

Section A. DC Relay Characteristics

1. Study the specific relay given to you for this activity. Determine the coil connection terminals and the switch contacts. Manually close the armature of the relay while observing the physical change in contacts.
2. With the multifunction meter prepared to measure resistance, determine the coil resistance of the relay.

   Resistance = ______ ohms.
3. Manually close the armature of the relay and determine the type of switch employed by the relay.

   Switch type is ____________.
4. Connect the circuit of Fig. 7-15A to determine the "pull-in" and "drop-out" characteristics of the relay. As a rule, you can see the physical change or hear a clicking sound when "pull-in" occurs.
5. Adjust the ammeter to the 500-mA range initially. Then adjust the variable dc source to zero volts and turn on the power supply. Gradually increase the voltage source until "pull-in" occurs. If possible, change the ammeter to a lower range to obtain a more accurate reading of the "pull-in" current value.
6. Reduce the voltage source to zero and test the "pull-in" current several times for accuracy.

   "Pull-in" current = ______ milliamperes.
7. Prepare the multifunction meter to measure dc voltage and determine the precise voltage where "pull-in" occurs.

   "Pull-in" voltage = ______ volts.
8. With the relay in its "pull-in" state, gradually reduce the voltage of the source to determine the armature "drop-out" current value. You may want to make two or three trial tests before deciding on a specific value.

   "Drop-out" current = ______ milliamperes.

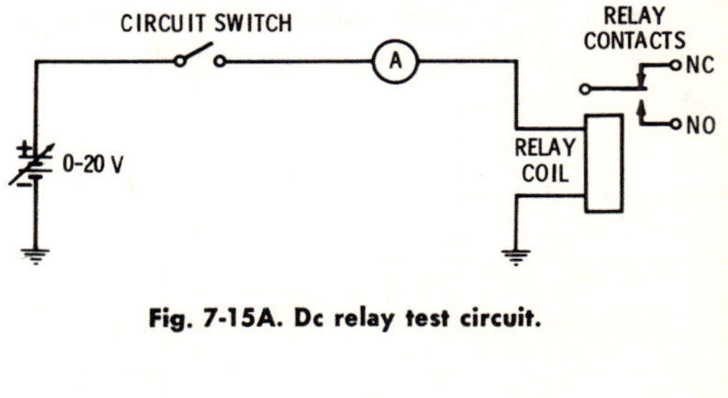

Fig. 7-15A. Dc relay test circuit.

9. Prepare the multifunction meter to measure resistance and test the continuity of the relay switch when it is energized by the dc source. The contacts of the relay are normally isolated from the relay coil.
10. Connect the relay switch contacts to energize one lamp from an independent 6.3-volt ac source when the relay armature is in its "pull-in" state. When the relay drops out, the lamp should turn off. Make a schematic diagram of the circuit design in the following space:

11. If time permits, alter the lamp switching circuit so that the lamp is "on" when the relay is in the "drop-out" state and goes "off" when it is in the "pull-in" state. Make a schematic diagram of the circuit in the space previously provided. The relay switch contacts are generally labelled "normally closed" (nc) or "normally open" (no) according to their state. Label the contacts of the switch on the schematic diagram.

Analysis

1. What would be the significance of knowing the "pull-in" and "drop-out" current characteristics of a relay?

2. In the manufacture of a relay, what determines the "pull-in" current?

3. How could a 125-ohm, 12-volt dc relay be used in a 30-volt dc circuit? What circuit modifications would be required?

4. Why are the contacts of a relay often included in the characteristic rating of a relay?

Section B. AC Relay Operation

1. Connect the ac relay test circuit of Fig. 7-15B.

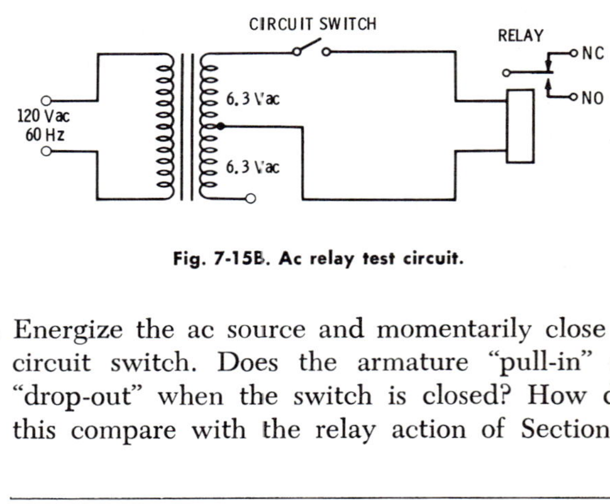

Fig. 7-15B. Ac relay test circuit.

2. Energize the ac source and momentarily close the circuit switch. Does the armature "pull-in" and "drop-out" when the switch is closed? How does this compare with the relay action of Section A?

3. Alter the circuit of Fig. 7-15B to conform with the circuit of Fig. 7-15C.

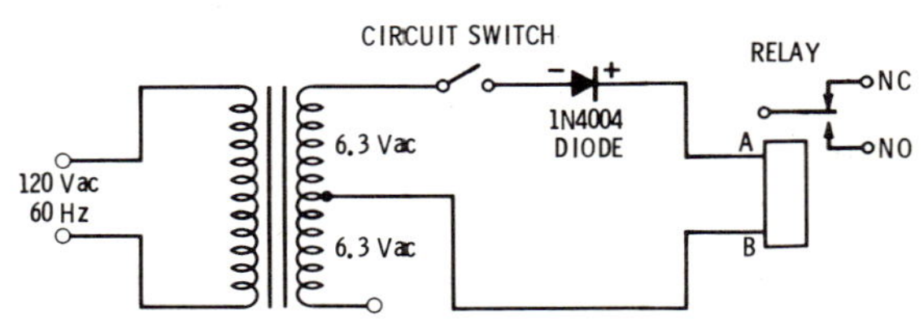

Fig. 7-15C. Ac relay test circuit with diode added.

4. Energize the ac source and momentarily close the circuit switch. Does the armature "pull-in" and "drop-out" when the switch is closed? How does it compare with the relay action of Step 2?

5. Connect a 4-μF capacitor across the relay coil with the positive lead to terminal A and the negative lead to terminal B.
6. Momentarily turn on the circuit switch and test the operation of the circuit. Does the armature "pull-in" and "drop-out"? How does the operation of this circuit compare with Steps 2 and 4?

7. Turn off the ac source and disconnect the circuit. Return all parts to the storage cabinet.

Analysis

1. Why does a hum occur when ac is applied to the relay of this part of the activity?

2. With the rectified dc circuit of Fig. 7-15C, why does hum still occur?

3. What is the function of the capacitor in Step 5?

Section C. Relay Latching Circuits

1. Using 12 volts dc as a source, design a relay circuit that will latch when it is energized initially by a "start" push-button or toggle switch. The relay should then remain energized until a second (or "stop") push-button or toggle switch is activated.
2. Make a schematic diagram of your relay latching circuit in the following space:

3. Explain the operation of your circuit to the instructor and get his approval before proceeding.

Instructor's Approval:

4. Construct the relay latching circuit and test its operation.
5. Alter the circuit so that it will turn on a No. 47 dial lamp in the latched state, and then turn the lamp on in the unlatched state.
6. Ask the instructor to approve of your lamp control circuit when it is completed.

Instructor's Approval:

7. Turn off the power source and disconnect your circuit. Return all parts to the storage cabinet.

Analysis

1. Where would a latching type of relay control circuit be used?

2. What kind of system control device is a relay?

3. After the relay is latched, why must a second switch be used to turn off the latching action?

7-16

Reed Switches

Objective

Reed switches are unique devices that respond to a controlled magnetic field. They are designed to make and break contact when exposed to either a permanent magnetic field or to an electromagnetic field.

The contacts of a reed switch are housed inside of a hermetically sealed glass tube. When actuated, contact sparks are isolated from the outside environment. Construction of this type lends itself well to explosion-proof applications. Switch contacts are also isolated from outside dust and corrosion, which means improved operating life expectancy.

Physically, a dry reed switch contains two flat metal strips, or "reeds," housed in a hollow glass tube filled with an inert gas. When the reeds are exposed to a magnetic field they are forced together, thus making or breaking contact depending on their design. The normally closed switch breaks contact when influenced by a field. Normally open contacts, by comparison, are forced closed when influenced by a magnetic field.

In this activity, you will construct a simple electrical circuit and control its action by changes in magnetic field strength. Through this experience you will gain some insight into the operation, sensitivity, and control action of a reed switch. Industrial applications of this device are quite numerous in explosive areas and dirt-prone environments.

Equipment

Dc power supply—0–5 V, 1.0 A.
Ac power source—6.3 V, 60 Hz.
Reed switch—GE-X7 or equivalent.
Bobbin-wound, reed-switch coil—General Electric.
Reed-switch magnet.
Lamp—No. 47.
Switch—spst.
Circuit construction board.

Procedure

1. Construct the reed-switch circuit of Fig. 7-16A.
2. Turn on the circuit switch and move the small permanent magnet near the reed switch. Try different combinations such as magnet at a right angle with reeds, parallel with reeds, etc. Which type of magnet orientation produces the best control capability?

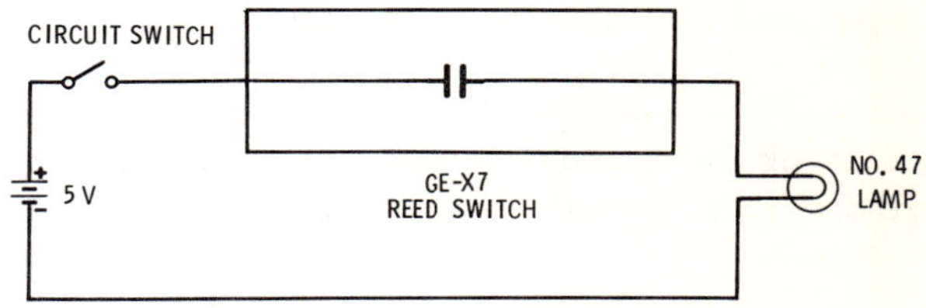

Fig. 7-16A. Reed-switch test circuit for observing permanent-magnet actuation.

____

3. Place the reed switch in the center of the bobbin-wound coil. Connect the circuit of Fig. 7-16B to test the electromagnetic action of the reed switch.
4. Turn on the ac power source and the dc coil-control circuit.

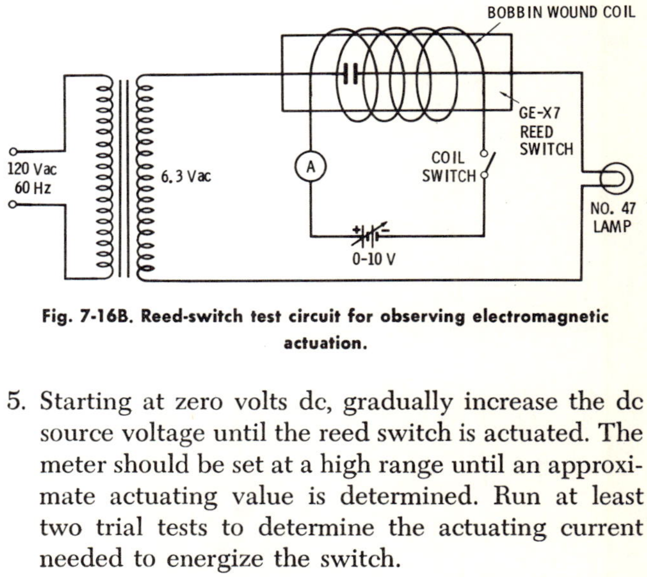

Fig. 7-16B. Reed-switch test circuit for observing electromagnetic actuation.

5. Starting at zero volts dc, gradually increase the dc source voltage until the reed switch is actuated. The meter should be set at a high range until an approximate actuating value is determined. Run at least two trial tests to determine the actuating current needed to energize the switch.

   Dc actuating current = ______ milliamperes.
6. In some applications, the actuating coil is used to increase the sensitivity of the reed switch by producing a partial field. Increase the dc coil current close to the actuating value. Then place the permanent magnet near the coil to actuate the reed. In this case, the switch can have some degree of variable sensitivity.

7. Turn off the ac and dc power sources for the circuit of Fig. 7-16B. Use 5 volts dc to supply the coil. The primary side of the coil should be connected to a variable ac transformer.
8. Turn on both the ac and dc sources, then increase the ac voltage applied to the coil. Measure the applied ac. Do not increase the voltage to a value that will cause the reed coil to overheat. How does the reed switch respond to ac compared with dc?

____

9. Turn off the ac and dc power sources and disconnect the circuit. Return all components to the storage cabinet.

Analysis

1. How could a reed switch be used to control a pump motor in a liquid sump tank?

____

____

____

2. Make a sketch of this circuit.

3. How could a reed switch be used to control a circuit in an explosive area without danger of explosion?

____

____

____

7-17

Synchro Systems

Objective

Synchro systems are commonly classified as two or more generator/motor units connected in such a way that they permit the transmission of angular shaft changes by electromagnetic field changes. When the shaft of the generator unit is turned, it automatically causes a corresponding change in the motor shaft position. Through systems of this type, it is possible to achieve accurate control over a distance.

Typical synchro systems contain two or more electromagnetic devices that are similar in appearance to a small electric motor. These devices are, however, connected in such a way that the angular position of the generator shaft is transmitted through interconnecting wires to the motor or receiving unit.

When electrical energy is first applied to a generator or motor unit, the unit will move to a specific location, then come to rest. This position is called electrical 0°. Position indicators usually point to system 0° in reference to electrical 0°. Both synchro generators and motors are adjusted to 0° in the same manner.

In this activity, you will have an opportunity to connect the motor and generator units of a simple synchro system together to test their operation. Through this experience, you will become familiar with actual circuit terminal designations and be able to connect the units together to achieve clockwise and counterclockwise rotation.

Equipment

Isolated variable ac power source—0–120 V @ 3 A.
Synchro generator/motor set.
Multifunction meter.
Oscilloscope.

Procedure

Section A. Electrical 0° Determination

1. Position the generator part of the synchro system so that you can readily observe the terminal connections at the rear of the unit. The stator terminals are usually labeled S_1, S_2, and S_3. The rotor terminals are generally labeled R_1 and R_2.
2. Connect the isolated variable ac power source to the rotor terminals as indicated in Fig. 7-17A.

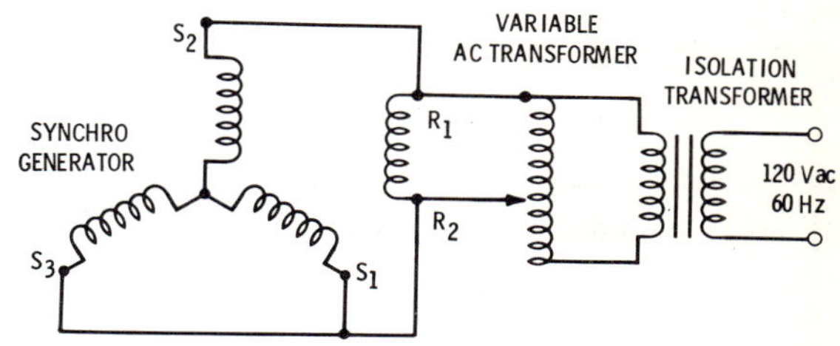

Fig. 7-17A. Circuit for determining the electrical 0° position of a synchro generator unit.

3. Adjust the variable ac transformer to its lowest output voltage level. Then turn on the ac source and gradually increase the voltage until the rotor winding locks in place. This represents the electrical 0° position of the rotor.
4. Prepare the multifunction meter to measure ac voltage and connect it across R_1 and R_2.

   The supply voltage measures ______ volts ac.
5. With the rotor locked at electrical 0°, adjust the indicating dial to the mechanical 0° position.
6. Another way to determine the electrical 0° position of a synchro generator is by the voltmeter method. Turn off the power source and connect the generator unit so that S_1, S_2, and S_3 are open with respect to the ac source applied to the rotor. The stator windings in this case are energized entirely by induction from the rotor.
7. Turn on the power source and measure the voltage across S_1 and S_2. Turn the rotor while observing the voltage. Note that zero voltage occurs at two locations during one revolution. This represents electrical 0° and 180°.
8. To distinguish between 0° and 180°, turn off the power source and connect a jumper wire between R_1 and S_2. Then turn on the power source and measure the voltage across R_2 and S_1. Turn the shaft or dial and note that the voltage has a high value at one position and a lower value at an opposite position. Electrical 0° and the low point are the same. The 180° position is represented by the highest voltage value.
9. Turn off the power source and disconnect the generator unit.
10. Connect the motor or receiver unit in the same manner outlined for the generator, and set its electrical 0° indicator.

11. Turn off the power source and disconnect the circuit. Return all parts to the storage cabinet.

Analysis

1. Explain how the stator windings are energized by inductance as noted in Step 6.

2. What is meant by the term electrical 0°?

3. What is the electrical difference between a synchro generator and a synchro motor of the same set?

Section B. Synchro Position Control

1. Connect the synchro generator/motor units together as indicated in Fig. 7-17B.

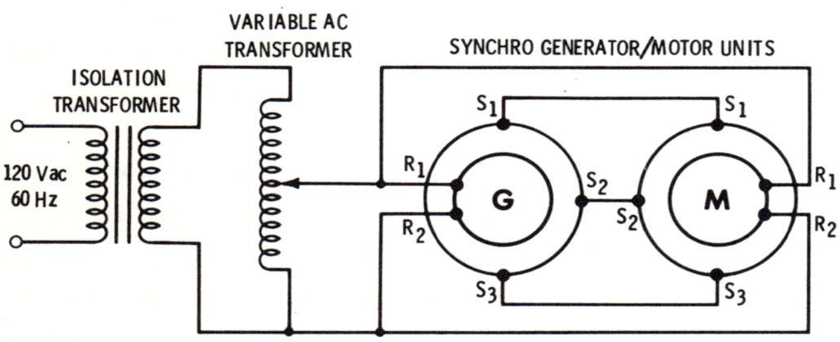

Fig. 7-17B. Circuit for observing position control of a synchro generator/motor system.

2. Turn on the power source and increase the voltage of the variable ac transformer to the rated rotor voltage. As a general rule this is usually 120 volts ac.
3. Move the generator rotor to the 0° position. The motor unit should follow or track the generator.
4. Rotate the generator 90° in the clockwise direction. How does the motor respond?

   Return the generator to the 0° position and then rotate the generator 90° counterclockwise. How does the motor respond?

   You may want to try a variety of position changes to confirm your answer.

5. Prepare the oscilloscope for operation and connect the ground lead to R_2 and the probe lead to R_1. Turn on the power source and turn the rotor through one complete revolution in each direction. How does the waveform respond to this physical change in the rotor?

6. Turn off the power source and connect the ground lead of the oscilloscope to S_1 and the probe lead to S_3. Turn on the ac power source and turn the rotor one revolution in each direction while observing the waveform. How does it respond?

7. Turn off the power source and reverse the R_1 and R_2 rotor leads of the motor unit. Turn on the power source and test the operation. How has the circuit changed compared with previous steps?

8. Turn off the power source and return R_1 and R_2 as they were originally connected in Fig. 7-17B. Connect stator leads S_1 of the generator to S_3 of the motor and S_3 of the generator to S_1 of the motor.
9. Turn on the power source and test the rotary control action of the generator/motor set. How does it respond when connected in this manner?

10. Turn off the power source and disconnect the circuit. Return all components to the storage cabinet.

Analysis

1. What does the waveform of Step 6 show about the stator voltage?

2. Why does the motor rotor track the generator rotor of a synchro system?

AC Synchronous Motors

Objective

The ac synchronous motor used in industrial processing equipment is commonly classified as a constant-speed device. It has extremely rapid starting, stopping, and reversing characteristics. Motors of this type only necessitate a simple cw–stop–ccw rotary switch for control. The starting and running currents of a synchronous motor are identical, which is unique for ac motors. This characteristic means that a motor of this type is immune to a high inrush of current when direction changes occur. As a general rule, the synchronous motor can even be stalled without producing any damage to the motor.

In this activity, a simple ac synchronous motor test circuit will be constructed so that you may observe its starting, stopping, and current flow characteristics. Motors of this type are commonly used as drive mechanisms for work tables in machinery operations. Starting begins within 1½ cycles of the line frequency, and stopping occurs in five mechanical degrees of rotation. This motor represents a unique part of all servomechanisms used in precision rotary control applications today.

Equipment

Ac synchronous motor—Superior Electric Type SS150 or equivalent.
Capacitor—3.75 μF, 330 V ac.
Resistor—250 Ω, 25 W.
Spst switches (3).
Isolated ac power source—120 V.
Ac ammeter—0–5 A.

Procedure

1. Construct the ac synchronous motor test circuit of Fig. 7-18A.
2. Turn on the cw switch for clockwise rotation.
3. Momentarily turn on the power switch and observe the rotation of the motor.
4. With a tachometer measure the speed of rotation.

   Speed = ______ r/min.

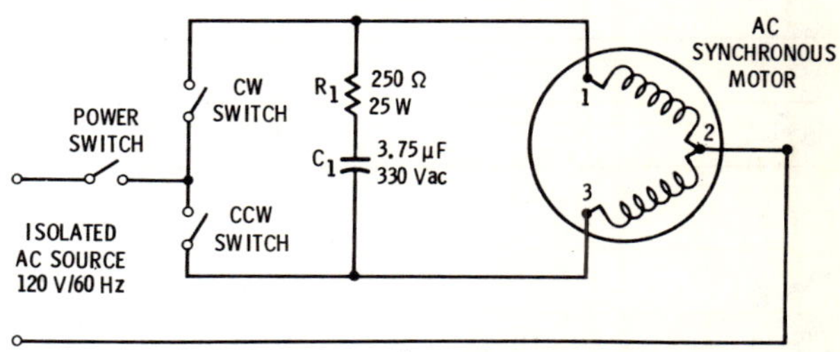

Fig. 7-18A. Ac synchronous motor test circuit.

5. Momentarily turn off the power switch and observe the stopping action of the synchronous motor. You may want to run several trials to see the quickness of the stopping action. How does this compare with other ac motors?

   ____

   ____

6. Note each time that the motor is turned on, how quickly it comes to speed.
7. With an ac ammeter, measure the running current and starting current of the synchronous motor.

   Running current = ______ amperes.

   Starting current = ______ amperes.
8. Holding down on the motor, wedge a piece of wood between the rotating shaft and the bench. This will provide a simple loading method for test purposes.
9. When the motor is loaded down, how does the running current respond?

   ____

10. If the motor is completely stalled, how does the running current respond?

    ____

11. Remove the load from the motor, turn off the cw switch, and turn on the ccw switch. The motor should rotate equally as well in the ccw direction and have the same basic characteristics. Test these again to verify the theory.
12. Switch off the ccw switch, then switch on the cw switch and notice the ease with which direction change occurs.

13. Turn off the ac power source and disconnect the circuit. Return all parts to the storage cabinet.

Analysis

1. What is the function of the resistor and capacitor of this test circuit?

2. What is an application of the ac synchronous motor?

3. How would the operation of an ac synchronous motor compare with the operation of a typical ac induction motor with respect to starting, stopping, and current flow?

7-19

DC Stepping Motors

Objective

Dc stepping motors represent a unique electromechanical rotary actuator that is used in automatic processing machinery. These motors are primarily designed to change electrical pulses into rotary motion. The amount of rotary movement or angular displacement produced by each pulse is repeated precisely for each succeeding pulse. The resulting rotation of these motors is primarily used to accurately locate or position work tables or fixtures for automatic machining operations. Very precise degrees of accuracy can be achieved with these devices.

In this activity, you will construct a simple dc manual switching test circuit for a dc stepping motor. You will be able to produce very accurate shaft movements by different switching step combinations. Both the four-step and eight-step switching combinations will be tested. Through this activity, you will be able to observe the physical operation and test the accuracy of motor shaft rotation. The switching operation of this activity is achieved automatically by logic devices in actual operating circuits. Precise switching operations obviously improve the accuracy of this circuit.

Equipment

Dc stepping motor—Superior Electric Type M061-FD02 or equivalent.
Spst switches (4).
Resistor—50 Ω, 20 W.
Resistor—1000 Ω, ½W.
Dc power supply—0–25 V, 1 A.
Diodes—1N4004 (4).

Procedure

1. Construct the stepping motor test circuit of Fig. 7-19A.

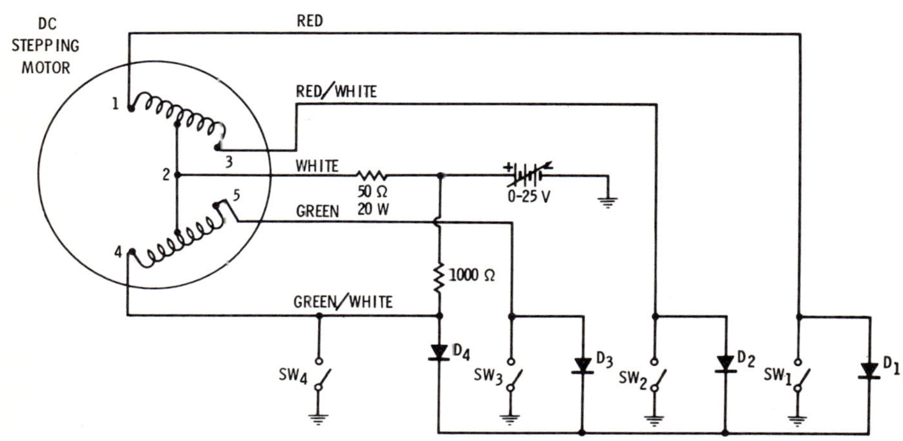

FOUR-STEP INPUT (FULL-STEP MODE)

STEP	SW_1	SW_2	SW_3	SW_4
1	ON	OFF	ON	OFF
2	ON	OFF	OFF	ON
3	OFF	ON	OFF	ON
4	OFF	ON	ON	OFF
1	ON	OFF	ON	OFF

EIGHT-STEP INPUT (HALF-STEP MODE)

STEP	SW_1	SW_2	SW_3	SW_4
1	ON	OFF	ON	OFF
2	ON	OFF	OFF	OFF
3	ON	OFF	OFF	ON
4	OFF	OFF	OFF	ON
5	OFF	ON	OFF	ON
6	OFF	ON	OFF	OFF
7	OFF	ON	ON	OFF
8	OFF	OFF	ON	OFF
1	ON	OFF	ON	OFF

Fig. 7-19A. Dc stepping motor test circuit.

2. The physical movement of the stepping motor shaft is small and rather difficult to observe. By attaching a simple indicating device such as a small wire or paper pointer to the motor shaft, the mechanical rotation can be readily observed.
3. The switching sequence in the full-step mode will produce 1.8° steps for each switching combination.
4. Adjust the dc power supply to 15 volts. Set the indicator to a starting position and mark it as the starting reference. Then turn on the switch combinations for Step 1 of the full-step mode from the chart of Fig. 7-19A.
5. Switch to the next switching combination indicated by Step 2 of the chart. Each combination will produce a rotary step.
6. Follow the stepping procedure through at least two of the four-step cycles. Notice that the switching step combination repeats itself after Step 4.
7. Mark the indicator location as the stopping position for reference.
8. Calculate the rotational degrees by multiplying the number of steps by 1.8°.
9. If a protractor is available, measure the number of degrees of rotation between the starting indicator mark and the last switching location mark of the motor shaft.
10. Repeat the procedure outlined in Steps 2 through 9 of this activity for the eight-step sequence. In this stepping mode, each step produces 0.9° of rotation.
11. Disconnect the circuit and return all components to the storage cabinet.

Analysis

1. Basically, how does a dc stepping motor compare with the operation of a dc motor?

____

____

____

2. How many steps of the full-step mode are needed to produce one revolution?

____

3. How many steps of the half-step mode are needed to produce one revolution?

____

4. What is an industrial application of the dc stepping motor?

____

Conductor Tables

Table A1-1. Properties of Conductors

	Size (AWG or MCM)	Area (cmil)	Number Wires	Diameter Each Wire (in)	DC Resistance at 25°C, 77°F (Ω/1000 ft)	
					Copper	Aluminum
AWG Sizes	18	1620	Solid	0.0403	6.51	10.7
	16	2580	Solid	0.0508	4.10	6.72
	14	4110	Solid	0.0641	2.57	4.22
	12	6530	Solid	0.0808	1.62	2.66
	10	10,380	Solid	0.1019	1.018	1.67
	8	16,510	Solid	0.1285	0.6404	1.05
	6	26,240	7	0.0612	0.410	0.674
	4	41,740	7	0.0772	0.259	0.424
	3	52,620	7	0.0867	0.205	0.336
	2	66,360	7	0.0974	0.162	0.266
	1	83,690	19	0.0664	0.129	0.211
	0	105,600	19	0.0745	0.102	0.168
	00	133,100	19	0.0837	0.0811	0.133
	000	167,800	19	0.0940	0.0642	0.105
	0000	211,600	19	0.1055	0.0509	0.0836
MCM Sizes	250	250,000	37	0.0822	0.0431	0.0708
	300	300,000	37	0.0900	0.0360	0.0590
	350	350,000	37	0.0973	0.0308	0.0505
	400	400,000	37	0.1040	0.0270	0.0442
	500	500,000	37	0.1162	0.0216	0.0354
	600	600,000	61	0.0992	0.0180	0.0295
	700	700,000	61	0.1071	0.0154	0.0253
	750	750,000	61	0.1109	0.0144	0.0236
	800	800,000	61	0.1145	0.0135	0.0221
	900	900,000	61	0.1215	0.0120	0.0197
	1000	1,000,000	61	0.1280	0.0108	0.0177

Table A1-2. Allowable Ampacities of Insulated Conductors; Single Conductor in Free Air (in Amperes)

AWG or MCM	Copper		Aluminum	
	60°C, 140°F R, T, TW	75°C, 167°F RH, RHW, TH, THW	60°C, 140°F R, T, TW	75°C, 167°F RH, RHW, TH, THW
14	20	20	Not	Used
12	25	25	20	20
10	40	40	30	30
8	55	65	45	55
6	80	95	60	75
4	105	125	80	100
3	120	145	95	115
2	140	170	110	135
1	165	195	130	155
0	195	230	150	180
00	225	265	175	210
000	260	310	200	240
0000	300	360	230	280
250	340	405	265	315
300	375	445	290	350
350	420	505	330	395
400	455	545	355	425
500	515	620	405	485
600	575	690	455	545
700	630	755	500	595
750	655	785	515	620
800	680	815	535	645
900	730	870	580	700
1000	780	935	625	750

APPENDIX **2**

Insulation Table

Type	Type Letter	Maximum Operating Temperature (°F)	Application
Rubber	R	140	Dry locations
Heat-resistant rubber	RH	167	Dry locations
Heat-resistant rubber	RHH	194	Dry locations
Moisture- and heat-resistant rubber	RHW	167	Dry and wet locations
Thermoplastic	T	140	Dry locations
Moisture- and heat-resistant thermoplastic	THWN	167	Dry and wet locations

APPENDIX **3**

Thermocouple Temperature-Voltage Curves

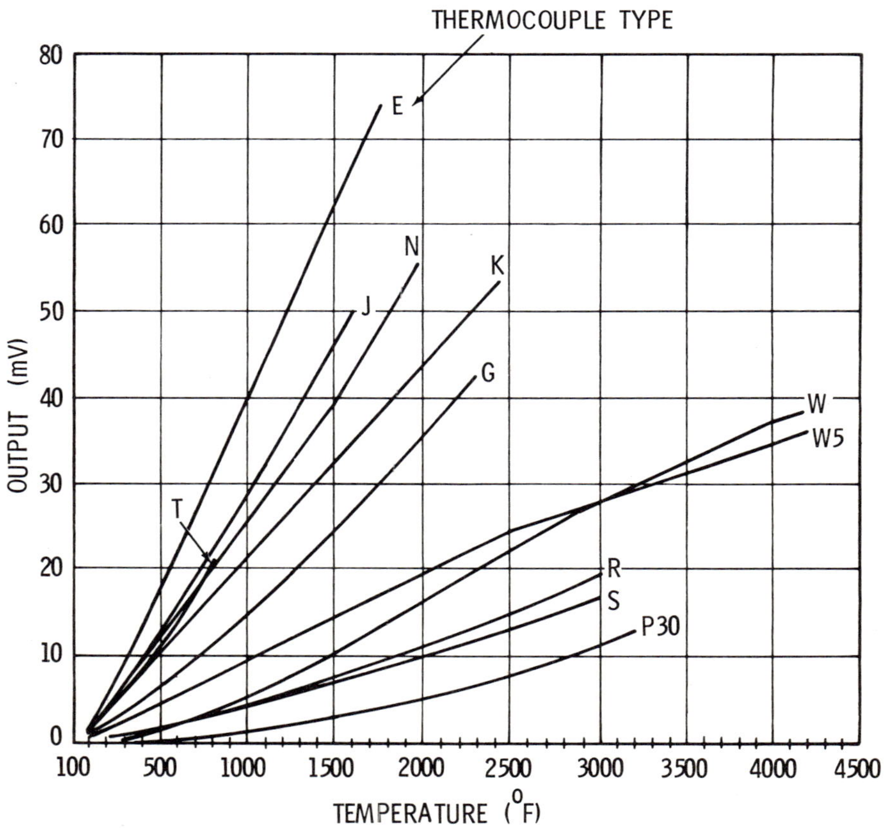